HANDBOOK OF ATTACHMENT INTERVENTIONS

HANDBOOK OF ATTACHMENT INTERVENTIONS

Edited by

Terry M. Levy

Evergreen Consultants in Human Behavior
Evergreen, Colorado

ACADEMIC PRESS

San Diego London Boston New York Sydney Tokyo Toronto

This book is printed on acid-free paper. 

Academic Press
An imprint of Elsevier Science
525 B Street, Suite 1900, San Diego, California 92101-4495, USA
http://www.academicpress.com

Academic Press
84 Theobald's Road, London WC1X 8RR, UK
http://www.academicpress.com

Library of Congress Catalog Card Number: 99-61960

International Standard Book Number: 0-12-445860-2

PRINTED IN THE UNITED STATES OF AMERICA
02 03 04 05 06 07 9 8 7 6 5 4

CONTENTS

2 Treating ADHD as Attachment Deficit Hyperactivity Disorder

RANDALL D. LADNIER AND ALICE E. MASSANARI

3 Parenting Children with Attachment Disorders

NANCY L. THOMAS

4 Integrating Attachment Concepts from Western Psychological and Buddhist Perspectives

BONNIE RABER WICKES

5 Problematic Attachment and Interrupted Development of Relationships: Causes, Interventions, and Resources within a Military Environment

RONALD G. BALLENGER

6 Permanency Planning and Attachment: A Guide for Agency Practice

DIANE SABATINE STEWARD AND KATHRYN ROSE O'DAY

7 Attachment Theory: A Map for Couples Therapy

SUSAN JOHNSON AND ANN SIMS

CONTRIBUTORS

Numbers in parentheses indicate the pages on which the authors' contributions begin.

John F. Alston (193) Private Practice, Evergreen, Colorado 80437

Ronald G. Ballenger (129) Educational and Developmental Intervention Services, 235th BSB CMR 463, Army Post Office, AE 09177

Susan Johnson (169) Department of Psychology, University of Ottawa, Ottawa, Canada K1N 6N5

Randall D. Ladnier (27) Family Counseling Center of Sarasota County, Inc., Sarasota, Florida 34239

Terry M. Levy (1, 243) Evergreen Consultants in Human Behavior, Suite 206, Evergreen, Colorado 80439

Alice E. Massanari (27) Family Counseling Center of Sarasota County, Inc., Sarasota, Florida 34239

Kathryn Rose O'Day (147) Children's Home Society of Florida, Ft. Lauderdale, Florida 33301-1094

Michael Orlans (1, 243) Evergreen Consultants in Human Behavior, Evergreen, Colorado 80439

Paula Pickle (261) The Attachment Center at Evergreen, Inc., Evergreen, Colorado 80437

Ann Sims (169) Department of Psychology, University of Ottawa, Ottawa, Canada K1N 6N5

Diane Sabatine Steward (147) Children's Home Society of Florida, Ft. Lauderdale, Florida 33301

Nancy L. Thomas (67) Therapeutic Parenting Specialist, Glenwood Springs, Colorado 81602

Bonnie Raber Wickes (111) Private Practice, Anchorage, Alaska 99501

PREFACE

As anyone who has ever planted a garden knows, you must first prepare the soil—make the soil fertile in order to foster health and growth. The same is true for children; they must have a context that promotes healthy functioning and development. *Attachment* between child and caregiver(s) is a major aspect of this crucial context. It is as basic as food and water, necessary for healthy development of the body, mind, relationships, values, and spirit.

Much has been written over the past 50 years about attachment theory, research, and clinical implications. Currently, interest in attachment is on the increase, as more and more mental health and child welfare professionals become aware of the significance of this issue regarding function and dysfunction in children, families, and society. *Handbook of Attachment Interventions* discusses, in depth, attachment-related topics, problems, and solutions that are important, not typically covered in detail in other volumes, and deserving of greater consideration. Thus, a primary goal of this book is to "widen the lens," to appreciate and understand the depth and breadth of attachment-related issues and solutions that are diverse yet extremely relevant in our current social climate.

What exactly is this "fertile soil" for children alluded to above? Another goal of this book is to answer that question in reference to varied attachment-related issues, populations, and sociocultural arenas. Thus, the chapters in this book focus on attentional deficits, hyperactivity, bipolar disorder, and other psychobiological conditions in children; aggression, violence, and antisocial personality; parenting children with severely disrupted attachment; integrating Western and Eastern perspectives; foster care, adoptive family, and military family systems; community agency, social service, and other child welfare systems; and adult individual and relationship functioning and treatment. An additional goal of this book is to provide a voice for the ideas, perspectives, and solutions found in the subsequent pages. Each chapter focuses on both theory and practice, with the emphasis on specific strategies and methodologies of intervention. The final goal of this book is to offer specific solutions to various attachment-related problems in children, adults, families, and social systems.

This book is intended primarily for mental health, social service, and allied child welfare professionals who are responsible for influencing the lives of children, adults, and families and for determining the moral framework of our society. Clinicians, caseworkers, therapeutic foster parents, junvenile and family court judges, and others who work in these realms will find interesting concepts and effective strategies for positive change in these chapters. Graduate students in child welfare, child development, and family systems, as well as adoptive parents and others interested in salient attachment-related topics, will find these contributions illuminating and useful.

I extend my appreciation to the authors in this volume. I am aware that each of these contributors is extremely busy and involved in his or her particular area of expertise. They took time from their work and tight schedules to complete these chapters. This reflects their high level of dedication and commitment to the issues and solutions discussed herein.

Terry M. Levy

1

ATTACHMENT DISORDER AS AN ANTECEDENT TO VIOLENCE AND ANTISOCIAL PATTERNS IN CHILDREN

TERRY M. LEVY
MICHAEL ORLANS

Evergreen Consultants in Human Behavior
Evergreen, Colorado

The United States is the most violent country in the industrialized world—particularly for children. Homicide is the 11th leading cause of death for all Americans, but the third leading cause of death for children between the ages of 5 and 14 (J. D. Osofsky, 1995). The homicide rate for young males is 40 times higher than the country with the lowest rate (Japan). Children and youths are victimized more than adults in every category—physical abuse, assault, bullying, rape. There was a 300% increase between 1986 and 1993 in the number of children seriously injured by maltreatment—mostly by violent parents (Children's Defense Fund, 1997).

The proliferation of violence has been likened to a national epidemic, breeding more violence at an exponential rate (Levine, 1996). A quarter of all households in the U.S. are victimized by crime each year. Nearly 1 million teenagers are victims of violent crime annually, with African-American males and those living in poverty at greatest risk. Even schools cannot provide a safe haven. Three million crimes occur on or near school grounds each year. One hundred and five fatalities were reported from 1992 to 1994; 40 children have been killed in school each year since 1993 (Kachur, Stennies, Powell, & Lowery, 1994).

Handbook of Attachment Interventions

Recent reports in the media of declining adult crime rates are misleading because they do not reflect the alarming increases in youth crime. There is a rapidly growing percentage of extremely cruel and violent crimes committed by children under the age of 12. We are experiencing a pace of violence among certain children that has been steadily rising since the mid-1970s. Violent crime among juveniles has quadrupled since 1975. A small percentage of disturbed youths are committing a larger percentage of violent crimes and at a younger age. Between 1983 and 1992, the arrest rate for girls under the age of 18 increased by 85%, while for boys it went up by 50%. The number of youths held in juvenile facilities has increased 41% since the late 1980s. More than 110,000 children under age 13 were arrested for felonies in 1994; 12,000 were crimes against people, including murder, rape, robbery, and aggravated assault (Berman, Kurtines, Silverman, & Serafini, 1996).

The children committing these violent crimes have a history of chronic aggression. Research shows that by age 4 they display ongoing and consistent patterns of aggression, rage, bullying, defiance, and controlling interactions with others (Greenberg, Deklyen, Speltz, & Endriga, 1997). Karr-Morse and Wiley (1997, p. 6) provide the following recent examples: A 10-year-old boy killed a 9-month-old baby by kicking and hitting her; a 4-year-old stomped an 8-week-old infant to death in her crib; a 10-year-old killed an 84-year-old neighbor by beating her with her cane; four 2nd-grade boys tried to kill a female classmate for breaking up with an 8-year-old gang leader.

Children are becoming more violent. Every 5 minutes a child is arrested for a violent crime. Juvenile homicide has doubled since the late 1980s (Children's Defense Fund, 1997). Violent crimes perpetrated by youths have recently become the focus of increased national concern and scrutiny. In an 8-month period (October, 1997–May, 1998), school shootings have resulted in 14 killed and 49 wounded. In West Paducah, Kentucky, a 14-year-old boy sprayed bullets into a high school prayer circle, killing three girls and wounding five others. In Pearl, Mississippi, a 16-year-old stabbed and killed his mother, then shot and killed two schoolmates and wounded seven other high school students. Two boys, ages 11 and 13, in Jonesboro, Arkansas, shot and killed four students and one teacher and wounded 10 others. On May 21, 1998, a 15-year-old boy killed two students and wounded 25 others in Springfield, Oregon, when opening fire in the school cafeteria—after he shot and killed his mother and father in their home.

This series of recent killings and injuries by angry youths marks a shift in the nature of youth violence, with incidents moving beyond one-to-one disputes into movie-style scenes of mass mayhem. In essence, there is more firepower, more victims, and an increased sense of callousness and indifference on the part of the young killers (Lewin, 1998). The vast majority of these young offenders have histories of abuse and neglect, lived in single-parent homes with young and highly stressed caregivers, and had parents with criminal records. Most of these children suffer from undiagnosed, untreated, and severe attachment disorders. They go on to commit more numerous and serious offenses, becoming the superpredators of tomorrow (Levy & Orlans, 1998).

Children with a history of severe attachment disorder develop aggressive, controlling, and conduct disordered behaviors, which contribute to the development of an antisocial personality. As early as the latency years and preadolescence, these children exhibit a lack of conscience, self-gratification at the expense of others, lack of responsibility, dishonesty, and a blatant disregard for the rules and standards of family and society. Teenage boys who have experienced attachment difficulties early in life are three times more likely to commit violent crimes (Raine, 1993). Disruption of attachment during the crucial first 3 years of life can lead to "affectionless psychopathy"; the inability to form meaningful emotional relationships, coupled with chronic anger, poor impulse control, and lack of remorse (Bowlby, 1969). These disturbing psychosocial qualities have contributed to a more violent and "heartless" character to the crimes being committed by today's youths.

CAUSES OF VIOLENCE IN CHILDREN

Prior to focusing on the association between disrupted attachment and violence, it is important to review other significant factors that contribute to aggressive and conduct-disordered behavior. A combination of emotional, social, and biological factors typically interact to promote later violence and antisocial acting-out (Levine, 1996; National Research Council, 1993). The interaction of internal vulnerabilities (e.g., emotional and/or cognitive deficits) and negative environmental factors (e.g., early abuse and/or neglect) create a context that results in youth violence. Extremely aggressive and violent juvenile offenders, for example, were found to have histories of maltreatment as well as cognitive/attentional/impulsivity problems (Lewis, 1990).

Family Influences

Common factors include parental mental illness, substance abuse, chronic discord and criminality, maternal rejection, depression and low I.Q., maltreatment, and multiple caregivers. Consider the following statistics on child maltreatment:

- The number of children seriously injured by maltreatment quadrupled from 1986 (140,000) to 1993 (600,000).

• Three million cases of maltreatment were investigated by Child Protective Services in 1995, and over 1 million were confirmed as serious abuse and/or neglect, with risk for continued maltreatment. Surveys indicated the actual number of cases were 10 to 16 times higher.

• Child Protective Services are unable to handle the vast increases; only 28% of seriously maltreated children were evaluated in 1993 compared to 45% in 1986 (Children's Defense Fund, 1997; Gallup, Moore, & Schussel, 1995; National Center on Child Abuse and Neglect [NCCAN], 1995).

Aggressive and violent children often have parents who have antisocial personalities, use harsh physical punishment, do not provide adequate supervision, and lack involvement in their children's lives. Severe family conflict and violence threatens childrens' fundamental security and leads to expectations and behaviors regarding violence (Emery & Laumann-Billings, 1998). Children who witness violence in their homes are at a high risk for developing distress symptoms (depression, anxiety, impulsivity, sleep problems) and violent behaviors (Martinez & Richters, 1993; Rosenberg & Rossman, 1990). Children who experience and/or witness violence in their families are seriously affected due to the literal and psychological proximity. Over 3 million children witness parental abuse each year, including physical abuse and fatal assaults. Domestic violence is associated with maltreatment of infants; mothers abused by their male partners have higher rates of child abuse. Physical abuse is the leading cause of death among children less than 1 year old (NCCAN, 1993; Strauss, 1993). Exposure to violence, including physical abuse, has severe and damaging consequences to many aspects of a child's functioning: physical, developmental, cognitive/attributional, social, emotional, behavioral, and academic (Kolko, 1996). Infants, toddlers, and older children often experience the three hallmark symptoms of Posttraumatic Stress Disorder: reexperiencing the traumatic event, numbing of responsiveness and avoidance of reminders of the trauma, and hyperarousal (American Psychiatric Association, 1994). Other common symptoms include sleep disturbance, night terrors, separation anxiety, fearfulness, aggressiveness, difficulty concentrating, and emotional detachment (Zeanah & Scheering, 1996).

Lack of parent–child involvement and supervision also plays a role. Three of four mothers of school-age children work outside the home. Juvenile crime is most common in the hours immediately after school, due to lack of supervision. Eighth-graders looking after themselves are more likely to smoke, drink, get poor grades, and use marijuana than children who have some supervision after school (Children's Defense Fund, 1997).

Environmental Factors

These include living in an impoverished atmosphere, modeling of violence in community, access to guns, and violence in the media. Human violence is largely learned. Children learn that violence is an acceptable way to solve problems by experiencing and witnessing violence (e.g., physical abuse, domestic violence). Boys who learn to be violent are more likely to be violent toward their wives and children and to be involved in crime when they become adults (Huesmann, Eron, Lefkowitz, & Walder, 1984). From preschool years through adulthood, violent individuals have thought patterns and belief systems that endorse the use of violence: "aggression is a legitimate way to express feelings, solve problems, boost self-image, and attain power." These thought patterns are usually learned in early childhood in the family and community (Shure & Spivack, 1988; Slaby & Guerra, 1988). Research on community violence revealed that in one Chicago neighborhood, one-third of school-age children witnessed a homicide and two-thirds witnessed a serious assault (Bell & Jenkins, 1993). Thirty-two percent of Washington, DC children and over half of New Orleans children were victims of violence in their community (Richters & Martinez, 1993). Children are directly exposed to family and community violence. Infants and toddlers, however, are indirectly but profoundly exposed; they are "tuned into" their caregivers' fears and anxieties about violence, influenced by the adults' coping strategies, and restricted in their psychosocial development (J. Osofsky, 1994).

The widespread availability and use of guns has broadened the scope and lethality of youth violence. One quarter of those arrested on weapons charges are juveniles (U.S. Department of Justice, 1995; cited in Levine, 1996). Firearms account for over 75% of all homicides for those 15 to 19 years old. Guns have become a staple of childhood and teenage life in many American cities. In one study, every child living in public housing in Chicago had witnessed a shooting by age 5. Every 90 minutes a child is killed by someone using a gun (Berkowitz, 1994; Mercy & Rosenberg, 1996).

The average American child spends 900 hours a year in school and 1500 hours a year watching TV. By the time a child leaves elementary school he or she has seen 8,000 murders and over 100,000 other acts of violence on television. Forty years of research has documented that violence is learned from TV and movies. Children's TV shows contain about 20–25 violent acts per hour. Preschoolers who watch violent cartoons are more likely to hit playmates, disobey class rules, and argue with teachers than are children who watch nonviolent shows. Elementary school children who watch considerable TV violence are more aggressive as teens and more likely to be arrested for criminal acts as adults. Children who watch extensive violence on TV can become less sensitive to the pain and suffering of others, more fearful in general, and more harmful to others. A primary message from TV and movies is that violence is an acceptable solution to human problems (Huston et al., 1992; National Institute of Mental Health, 1982).

Biological Contributions

These factors include prenatal drug and alcohol exposure, failure to thrive, severe and chronic maternal stress, birth-related complications and prematurity, nutritional deficiencies, and genetic background. There is no single "violence gene," but violence is related to traits that may be partially heritable—a difficult, fearless, and uninhibited temperament, hyperactivity, and attention problems. Temperament differences may partially explain why siblings are different and why, even in violent communities, only some youths turn to violence (National Research Council, 1993). Genetic deficits which result from prenatal drug and/or alcohol exposure, or other environmental causes, can lead to later violent behavior. The brain of the developing fetus can be damaged, resulting in problems with learning, attention, and impulse control later in childhood (Karr-Morse & Wiley, 1997; Schore, 1994).

Additional Factors

These include Attentional Deficit Hyperactivity Disorder (ADHD), low verbal I.Q., and posttraumatic stress symptoms. Children with ADHD are deficient in focusing, attending, planning, impulse control, concentrating, and self-regulating. Attentional Deficit Hyperactivity Disordered children who experience maltreatment and instability in their families often develop Oppositional Defiant Disorder (ODD) and Conduct Disorder (CD) over time. These children often become angry, defiant, and violent (Barkley, 1990). Children with symptoms of Post-traumatic Stress Disorder typically display aggressive and violent behaviors, with both a biological and emotional basis (Barkley, 1990; see pp. 22–23 for more details).

SECURE ATTACHMENT

Attachment is an enduring affective bond characterized by a tendency to seek and maintain proximity to a specific person, particularly when under stress (Ainsworth, 1973; Bowlby, 1969). Attachment is the deep and long-lasting emotional connection established between a child and caregiver in the first several years of life. It profoundly influences every component of the human condition—mind, body, emotions, relationships, and values. This is not something that parents do to their children; rather, it is something that children and parents create together in an ongoing reciprocal relationship. Attachment to a protective and loving caregiver who provides guidance and support as a basic human need is rooted in millions of years of evolution. There is an instinct to attach: babies instinctively reach out for the safety and security of the "secure base" with caregivers; parents instinctively protect and nurture their offspring. Attachment is a physiological, emotional, cognitive, and social phenomenon. Instinctual attachment behaviors in the baby are activated by cues or signals (i.e., social releasers) from caregivers (smile, eye contact, holding, rocking, touching, feeding). The attachment process is defined as a "mutual regulatory system," with the baby and caregiver influencing one another over time.

Beyond the basic function of secure attachment—providing safety and protection for the vulnerable young via closeness to a caregiver—there are several other important functions for children developmentally:

- to learn basic trust and reciprocity, which serves as a template for all future emotional relationships.
- to explore the environment with feelings of safety and security ("secure base"), which leads to healthy cognitive and social development.
- to develop the ability to self-regulate, which results in effective management of impulses and emotions.
- to create a foundation for the formation of identity, which includes a sense of competency, self-worth, and a balance between dependence and autonomy.
- to establish a prosocial moral framework, which involves empathy, compassion, and conscience.
- to generate the core belief system, which comprises cognitive appraisals of self, caregivers, others, and life in general.
- to provide a defense against stress and trauma, which incorporates resourcefulness and resilience (Levy & Orlans, 1995, 1998).

Children who begin their lives with secure attachment fare better in all aspects of functioning as development unfolds. Numerous longitudinal studies have demonstrated that securely attached infants and toddlers do better later in life regarding: self-esteem, independence and autonomy, enduring friendships, trust and intimacy, positive relationships with parents and other authority figures, impulse control, empathy and compassion, resilience in the face of adversity, school success, and future marital and family relations (Jacobson & Wille, 1986; Main, Kaplan, & Cassidy, 1985; Sroufe, Carlson, & Shulman, 1993; Troy & Sroufe, 1987; Waters, Wippman, & Sroufe, 1979).

Many professionals in the child welfare and mental health field focus on the symptoms associated with violence in youths (e.g., defiance, anger, impulsivity). The core issue of attachment, however, is often overlooked.

> In order to understand the tide of violent behavior in which America is now submerged, we must look before preadolescence, before grade school, before preschool, to the cradle of human formation in the first thirty-three months of life.

> These months, including nine months of prenatal development and the first two years after birth (33 months), harbor the seeds of violence for a growing percentage of American children (Karr-Morse & Wiley, 1997, p. 9).

The creation of a secure attachment relationship between the child and the primary caregiver(s) is a primary protective factor against later violent and antisocial patterns of cognition, behavior, and interaction. The specific attachment-related protective factors that reduce the likelihood of later aggression and violence include:

1. *Ability to regulate and modulate impulses and emotions:* Secure attachment with a primary caregiver is critical if children are to learn self-control. "The primary function of parents can be thought of as helping children modulate their arousal by attuned and well-time provision of playing, feeding, comforting, touching, looking, cleaning, and resting—in short, by teaching them skills that will gradually help them modulate their own arousal" (van der Kolk, 1996, p. 185).
2. *Developing prosocial values, empathy, and morality:* Considering the wants and needs of others was in our evolutionary interest. Sharing with the young, weak, and vulnerable made us more altruistic. This altruism became reciprocal, forming the evolutionary basis for "good behavior" (Morris, 1994). Secure attachments foster prosocial values and behaviors, including empathy, compassion, kindness, and morality.
3. *Establishing a solid and positive sense of self:* Children who experience a secure base with an appropriately responsive and available caregiver are more likely to be autonomous and independent as they develop. They are able to explore their environment with more confidence and less anxiety, resulting in enhanced self-esteem, feelings of mastery, and differentiation of self. They develop positive beliefs and expectations about themselves and relationships ("positive internal working model"). The route to caring for others begins with a positive and secure sense of self, with clear boundaries between self and others.
4. *Ability to effectively manage stress and adversity:* Resilience refers to an individual's competence and successful adaptation following exposure to significant adversity and stressful life events. Studies of resiliency have consistently found that the most basic and important protective factor is the history of caregiver–child attachment. Secure attachments are a primary defense against the development of severe psychopathology associated with adversity and trauma (Werner & Smith, 1992).
5. *Ability to create and maintain emotionally reciprocal relationships:* Securely attached children experience warm, trusting, and loving reciprocal relationships, resulting in the internalization of prosocial standards of behavior, cooperation, and self-control. They are able to experience, receive, and give affection and love. The child is "in sync" with the parent and, therefore, is learning to be aware of the feelings and needs of another

person. Secure attachment implies greater awareness of the mental state of others, which not only produces a more rapid and effective evolution of morality, but also protects the child from antisocial behavior.

DISORDERED ATTACHMENT

Infancy and the first several years of life is the critical developmental stage in which children develop basic trust, patterns of relating, sense of self, conscience, and cognitive abilities. Many children, however, do not experience secure attachments with loving, reliable, and protective caregivers and are left without a crucial foundation for healthy development. They are flooding our child welfare and juvenile justice systems with an overwhelming array of problems and are at a high risk for becoming adults who perpetuate the cycle with their own children. Children who begin their lives with seriously compromised and disrupted attachment often become impulsive, extremely oppositional, lacking in conscience and empathy, unable to give and receive genuine affection and love, angry, aggressive, and violent. A recent newspaper article reveals the seriousness of our current problem:

> The nation's juvenile courts, long a troubled backwater of the criminal justice system, has been so overwhelmed by the increase in violent teenage crime and the breakdown of the family that judges and politicians are debating a solution that was once unthinkable—abolishing the system and trying most minors as adults (Butterfield, 1997).

The legal and child welfare systems not only find it impossible to keep up with new cases, but have difficulty monitoring and serving the children and families on their current caseloads. As many as 50% of all fatalities that are due to child abuse and neglect occur in cases that have already been brought to the attention of law enforcement and child protection agencies (Lung & Daro, 1996).

The causes of attachment disorder are grouped into three categories: (1) parental/caregiver contributions (e.g., abuse and neglect, depression, psychological disorders); (2) child contributions (e.g., difficult temperament, prematurity, fetal alcohol syndrome); and (3) environmental contributions (e.g., poverty, stressful and violent home and/or community). The most common causes of attachment disorder are abuse, neglect, multiple out-of-home placements (e.g., moves in foster care system), and other prolonged separations from the primary attachment figure (e.g., hospitalization, prison, postpartum depression). Social service and mental health professionals often suggest that attached disorder is rare. The evidence, however, indicates otherwise. Research has shown that severe attachment disorders are created in up to 80% of children in high-risk families; risk factors include abuse, neglect, domestic violence, poverty, substance abuse, history of maltreatment in parents' childhoods, depression, and other serious psychological disorders of parents. Since there are 1 million substantiated cases of serious abuse and neglect in the U.S. each year, the statistics indicate there are approximately 800,000 children with severe attachment disorder coming to the attention of the child welfare system annually (Lyons-Ruth, 1996). Surveys indicate that the actual number of cases are 10 to 16 times higher (Gallup et al., 1995). These figures do not include the thousands of children with attachment disorder adopted from other countries.

Attachment disorder affects many aspects of a child's functioning. Symptoms exist along a continuum, from mild to severe, and are divided into six categories.

- *Behavior:* oppositional and defiant, impulsive, destructive, lie and steal, aggressive and abusive, hyperactive, self-destructive, cruel to animals, fire setting.
- *Emotions:* intense anger, depressed and hopeless, moody, fearful and anxious (although often hidden), irritable, inappropriate emotional reactions.
- *Thoughts:* negative core beliefs about self, relationships, and life in general ("negative working model"), lack of cause-and-effect thinking, attention and learning problems.
- *Relationships:* lacks trust, controlling ("bossy"), manipulative, does not give or receive genuine affection and love, indiscriminately affectionate with strangers, unstable peer relationships, blames other for own mistakes or problems, victimized others/victimized.
- *Physical:* poor hygiene, tactilely defensive, enuresis and encopresis, accident prone, high pain tolerance, genetic predisposition (e.g., depression, hyperactivity).
- *Moral/spiritual:* lack of empathy, faith, compassion, remorse, meaning and other prosocial values; identification with evil and the dark side of life.

Attachment disorder is transmitted intergenerationally; children with disordered attachments commonly grow up to be parents who are not able to create a secure foundation with their own children. Instead of following the instinct to protect, comfort, and love their children, they abuse, neglect, and abandon. There is a "pyramid effect"; with each generation there is a multifold increase in the number of children with attachment disorder.

Aggression and Antisocial Patterns

Neglectful, abusive, and nonresponsive caregivers produce out-of-control, angry, depressed, and hopeless children by 2 to 3 years of age. Attachment-disordered children have frequent and prolonged temper tantrums, are impulsive and accident prone, and desperately seek attention via negative behaviors. They are restless, irritable, have a brief attention span, demand instant gratification, and have little frustration tolerance by the preschool years. By age 5, they are angry, oppositional, and show lack of enthusiasm for learning. Their inability to control impulses and emotions leads to aggressive acting-out and lack of enduring relationships. Compared to securely attached children, attachment-disordered children are significantly more aggressive, disruptive, and antisocial.

It is the children with histories of disorganized-disoriented attachment who are most at risk for developing severe problems, including aggression. Disorganized attachment refers to a lack of, or collapse of, a consistent or organized strategy to respond to the need for comfort and security when under stress (Main & Solomon, 1990). Disorganized infant attachment has been found to be associated with unresolved loss, fear, and trauma of the parent(s). They have not mourned losses, are frightened by memories of past trauma, may dissociate, and script their child into unresolved family drama (Main & Goldwyn, 1984; Van IJzendoorn, 1995). Mothers of disorganized infants typically have histories of family violence and abuse rather than neglect alone (Lyons-Ruth, Alpern, & Repacholi, 1993). These mothers are "out of sync" with their babies, displaying confusing and mixed messages (e.g., extend arms toward infant while backing away) and inappropriate responses to their infant's cues (e.g., laugh when baby is in distress) (Lyons-Ruth, 1996; Main et al., 1985; Spieker & Booth, 1988). These mothers show high levels of negative and downcast affect to their babies and low levels of tenderness and affection (DeMulder & Radke-Yarrow, 1991). Thus, disorganized attachment is transmitted intergenerationally; parents raised in violent, frightening, and maltreating families transmit their fear and unresolved losses to their children through insensitive or abusive care, depression, and lack of love and affection. The infant is placed in an unresolvable paradox: closeness to the parent both increases the infant's fear and, simultaneously, need for soothing contact. Closeness and contact with the parent triggers fear rather than safety or comfort (Lyons-Ruth, 1996; Main & Hesse, 1990).

Kindergarten children who were classified as disorganized in infancy were six times more likely to be hostile and aggressive toward peers than were those classified as secure (Lyons-Ruth et al., 1993). Infants of impoverished adolescent mothers are at risk for developing severe attachment disorder and subsequent aggression. Sixty-two percent of these infants had disorganized attachment relationships and were more likely to initiate conflict with their mothers by aggressive and oppositional behavior by 2 years of age (Hann, Castino, Jarosinski, & Britton, 1991). These mothers were less affectionate and more rejecting of their child's overtures than other mothers. By the time they became toddlers, these children were aggressive, avoided and resisted their mothers, and were developing a controlling and coercive strategy to cope. Findings from the Minnesota High Risk Study, which followed a large community sample of impoverished mothers and infants from birth into adolescence, documented the relationship between insecure attachment and later conduct disorders. Insecurely attached infants were more aggressive and impulsive and had more conflict with peers and caregivers during their school years (Egeland, Pianta, & O'Brien, 1993; Erickson, Sroufe, & Egeland, 1985; Renken, Egeland, Marvinney, Mangelsdoft, & Sroufe, 1989; Sroufe, Egeland, & Kreutzer, 1990). Among high-risk families, it is clear that early disturbed attachment patterns place children at high risk for later aggression and violent behavior.

Children with a history of severe (disorganized) attachment disorder develop aggressive, controlling, and conduct-disordered behaviors, which contribute to the development of an antisocial personality. As early as the latency years, these children exhibit a lack of conscience, self-gratification at the expense of others, lack of reliability and responsibility, dishonesty, and a blatant disregard for the rules and standards of family and society. They are "macro-defiant"; they defy parental authority, social rules, millions of years of human evolution (i.e., evolutionary basis of moral behavior), and even reject the concept of a higher power. They exhibit a level of anger and noncompliance beyond that of typical Oppositional Defiant Disorder. The tendency to be controlling toward caregivers and others is a foremost symptom of disorganized attachment and a constant challenge for those who care for these children. They typically lie as a way of life. Pathological lying becomes a habitual strategy to avoid punishment and gain power and control ("I know the truth and you don't"). They often lie even when they do not have to, for no apparent purpose, which provides a sense of excitement and feeling of having the "upper hand." Symptoms and traits of the adult psychopathic and antisocial personality are displayed in severely attachment-disordered children: cruelty to animals; enuresis; fire-setting; predatory and vengeful; controlling and manipulative; lack of empathy, remorse and conscience; pathological lying, self-gratification at other's expense; inability to form close relationships (Hare, 1993; Yochelson & Samenow, 1993). Davis (1998) reports that serial killers seek control over others, lack conscience, and display other typical symptoms by age 12 (enuresis, animal torture, fire-setting, pathological lying, chronic daydreaming, violent fantasies). Douglas (1998) describes the adult violent psychopathic personality (e.g., sexual predator) as obsessed with manipulation, domination, and control. In essence, children with severe attachment disorders are violent psychopaths in training.

Children with histories of severe attachment disorder commonly display three behaviors that are also found in the childhood histories of adult psychopaths: cruelty to animals, enuresis, and fire-setting. Infamous serial murderers such as Jeffrey Dahmer, Ted Bundy, David Berkowitz ("Son of Sam"), Charles Manson, and Albert DeSalvo ("Boston Strangler") all tortured animals when they were young. Cruelty to animals is one of the most disturbing manifestations of attachment disorder. It ranges from annoyance of family pets (e.g., tail-pulling, rough play, kicking) to severe transgressions (e.g., strangulation, mutilation). These children lack the capacity to give and receive affection with pets, lack the motivation and sense of responsibility necessary to provide appropriate care, and are not able to empathize with the suffering of animals. They often delight in venting their frustrations and hostilities on helpless creatures to compensate for feelings of powerlessness and inferiority. Studies found that children who abuse animals are five times more likely to commit violent crimes as adults. The FBI's Behavioral Science Unit found that a majority of individuals who have committed multiple murders admitted to cruelty to animals during childhood (Cannon, 1997). Parental abuse of children was the most common etiological factor found in cruelty to animals (Tapia, 1971). Fromm (1973) noted that children who are sadistic are usually themselves the victims of cruel treatment. Showalter (1983) concluded that cruelty to animals represents a displacement of aggression from the child to a helpless animal.

Fire provides a particular appeal for severely attachment-disordered children. It's attributes of power and destruction are attractive qualities to the child who is rage filled and feels powerless. The child's fire-setting behaviors are extremely disconcerting to caregivers. The child senses this fear and apprehension and uses this to his or her advantage in order to gain further power and control. Fire-setting behaviors vary in degree from simple fascination and/or occasional lighting of matches to more serious actions such as setting fire to a home. The more serious the nature of the premeditated fire, the more seriously disturbed the child. Society is beginning to recognize the magnitude of this problem; juveniles now account for the majority of all arson arrests (55% in 1994). One-third of those arrested were under 15 years old, and 7% were under the age of 10. No other serious felony has such a high rate of juvenile involvement (Estrin, 1996).

Negative Working Model

The internal working model (core belief system) is the cognitive representation of early attachment relationships. Children develop beliefs about themselves, relationships, and life in general based on the nature of early attachment patterns with primary caregivers. This internal working model affects how the child interprets events, stores information in memory, and perceives social situations. A comparison of the internal working model of children with secure and disordered attachment follows:

Secure attachment

- *Self:* "I am good, wanted, worthwhile, competent, lovable."
- *Caregivers:* "They are appropriately responsive to my needs, sensitive, caring, trustworthy."
- *Life:* "The world is safe, life is worth living."

Disordered attachment

- *Self:* "I am bad, unwanted, worthless, helpless, unlovable."
- *Caregivers:* "They are unresponsive to my needs, insensitive, hurtful, untrustworthy."
- *Life:* "The world is unsafe, life is not worth living."

The internal working model of children with attachment disorder includes negative self-evaluations and self-contempt. They internalize lack of adequate care, love, and protection as self-blame and perceive themselves as unlovable, helpless, and responsible for maltreatment. Research has shown that this framework of negativity results in misinterpretation of social cues, including the tendency to attribute hostile intentions to others (Dodge, Bates, & Pettit, 1990). The child is conditioned to perceive threat and hostility even when it is not there and commonly responds with aggressive and coercive behavior (Troy & Sroufe, 1987). These latter core beliefs promote a sense of alienation from family and society; a need to control others and protect oneself at all times; and angry, vindictive, violent, and antisocial behaviors.

Lack of Morality and Antisocial Values

Important prosocial values, attitudes, and behaviors are learned in the context of secure attachment relationships via four psychological processes: (1) *modeling* by parents or other attachment figures, (2) *internalizing* the values and behavior of parents or other attachment figures, (3) experiencing *synchronicity* and *reciprocity* in early attachment relationships, and (4) developing a positive *sense of self.* When the family does not promote secure attachment and appropriate socialization experiences, as is the case with attachment disorder, the child is at risk for developing not only conduct disorders, but also a more pervasive lack of morality. Empathic parents rear empathic children. Children with compromised and disrupted attachments lack the models of empathy and compassion and tend to be cruel, controlling, and selfish. They have internalized antisocial values and standards, such as sadistic power, dishonesty, selfishness, and aggressive control. Their inner voice does not include a conscience or feelings of remorse. Lacking a secure foundation, a weak and negative self-identity has been created, and the child assumes a fearful and punitive orientation. There is no room for empathy or caring as the child must survive in a world perceived as lonely and threatening (Waters et al., 1979; Zahn-Waxler, Radke-Yarrow, Wagner, & Chapman, 1992).

Neurobiology of Attachment and Trauma

An extremely damaging and debilitating consequence of disordered attachment in children is their chronic inability to modulate emotions, behaviors, and impulses. This affects the biological and psychological ability to self-regulate and often leads to a variety of psychosocial problems, including aggression against the self and others (van der Kolk & Fisler, 1994). Secure attachment with a primary caregiver is critical if children are to learn self-control. Attachment and nurturing behaviors (eye contact, reciprocal smile, holding, rocking, touching) help to maintain the infant's homeostatic balance both emotionally and physically (Hofer, 1995; van der Kolk, 1996). This is a process that caregivers and babies accomplish together. This mutual regulatory process breaks down under conditions of anxious and disrupted attachment. Depressed, substance-abusing, or otherwise neglectful and abusive caregivers are not attuned to their infant's emotions and needs, leaving the baby without any necessary regulatory support (Robinson & Glaves, 1996).

The first 33 months (fetal stage and first 2 years) is the time of most rapid brain growth; the period when the quality of the caregiving environment profoundly affects the structure and function of the developing brain. Trauma affects the young on many levels of biological functioning. Threats to the infant and young child that are of sufficient intensity, duration, or frequency, such as abuse, neglect, and anxious-disorganized attachment, trigger an alarm reaction ("fight, flight, freeze"). This instinctual response to real or perceived danger is a normal response to acute stress. Traumatic experiences during infancy and childhood, however, can trigger prolonged alarm reactions, which alter the neurobiology of the brain and central nervous system. Lack of secure attachment and exposure to traumatic stress alters the nervous system, predisposing the child to be impulsive, overreactive, and violent (Perry, 1994, 1995). These children often develop symptoms of posttraumatic stress disorder, including: (1) recurring intrusive recollections, such as dreams or flashbacks; (2) persistent avoidance of stimuli associated with the trauma and numbing of general responsiveness; and (3) hyperarousal, such as hypervigilance, startle response, sleep difficulties, irritability, anxiety, and physical hyperactivity (Perry, 1994).

The limbic system is the part of the brain that controls emotion, impulses, and maternal functions. The cortex is involved in higher mental functions, such as logic and planning. The orbitofrontal cortex connects these two parts of the brain—a key area that is crucial to both attachment and emotional regulation. The orbitofrontal cortex blends input from the environment with visceral signals from inside the body and is especially sensitive to facial expressions (e.g., mother's smile). Compromised attachment, resulting from abusive and/or unresponsive caregiving, can inhibit the proper development of this brain system—the result is often impulsivity and violence (Schore, 1994, 1996; cited in Karr-Morse & Wiley, 1997, pp. 36–38). Bessel van der Kolk, a leading researcher in the field of trauma, reviewed the literature on the psychobiology of trauma and attachment. He writes, "Secure attachment bonds serve as a primary defense against trauma induced psychopathology ... the quality of the parental bond is probably the single most important determinant of long-term damage" (van der Kolk, 1996, p. 185).

TREATMENT

Traditional psychotherapeutic approaches are too often ineffective with severely attachment-disordered children, who do not trust or form a working alliance basic to success in therapy. Compromised attachment in the early years results in a need to control, fear of closeness, and a lack of reciprocity. The therapeutic challenge is to take charge in a firm yet caring way and gradually form a working relationship with the child. The same characteristics that make it difficult to help those with antisocial personality (no empathy or remorse, angry, defiant, dishonest, self-centered) are present in these children. The therapeutic challenge is to instill the basics—trust, empathy, cooperation, and conscience—qualities essential for successful living in a family and community. Through refining concepts and methods for the treatment of severely attachment-disordered children and families since the mid-1970s, we have found that effective therapy involves the following components.

- **Creating attachment patterns:** The primary therapeutic goal is to facilitate secure attachment in the parent–child relationship. To achieve this goal it is necessary to recreate the elements of secure attachment which were unavailable in the child's early developmental stages. In the context of the Holding Nurturing Process (HNP), children are provided with structure, attunement, empathy, positive affect, support, and reciprocity. The HNP is a therapeutic relationship and milieu which promotes secure attachment via social releasers, safe containment, corrective touch, access to "old brain" functions which control attachment behavior, and the development of a secure base in which positive developmental changes occur.
- **Systemic:** Attachment develops and is maintained in the context of overlapping relationship systems, including parent–child, marital, family, extended kin, and community. For example, it is common for attachment-disordered children to "triangulate" parents and other caregivers, playing one against the other. Effective treatment must address the various social systems in the life of the child and family.
- **Holistic and integrative:** Treatment focuses on mind, body, behaviors, emotions, relationships, and values. Therapeutic interventions and strategies are varied—experiential, psychoeducational, cognitive, skill-based. This approach is based on the concept that many factors interact to create both health and dysfunction.
- **Revisit, revise, revitalize:** Treatment is developmental, requiring the successful completion of each stage building upon the next. Attachment

trauma is first revisited to address core issues. Next, revisions are facilitated in belief systems, choices, relationship patterns, and coping skills. Last, revitalization includes celebrating achievements, cementing positive changes, creating plans, and enhancing hope for the future.

Many children with attachment disorder are adopted by well-meaning parents who are ill-prepared to handle their severe emotional and behavioral problems. These children are unable to give and receive love and affection, constantly defy rules and authority, are physically and emotionally abusive to caregivers and siblings, and create ongoing stress and turmoil in the family. As a result of insufficient preplacement services (education, training, support, matching) and postplacement services (individual and family therapy, parent education, support), family members and marriages suffer and the children do not improve. These parents have been "through the mill" of mental health and social services programs. They are commonly blamed for their child's problems, denied access to social service records, and thoroughly frustrated in their attempts to get help. They are angry with their child, feel guilty and inadequate, and are often on the verge of relinquishment. The therapeutic challenge is to enhance parents' motivation, positive emotion, and hope and encourage a more effective framework for conceptualizing their parenting role and relating to their child. Treatment for the child and parents focuses on the following.

1. **Child:** address trauma, attachment disorder, and negative working model (negative belief system and self-image); learn prosocial coping skills (communication, anger management, problem-solving), encourage respect, responsibility, resourcefulness, and reciprocity.
2. **Parent–child relationship:** enhance secure attachment patterns (trust, affection, intimacy, communication, reciprocity); reduce anger and negative patterns of relating.
3. **Family issues:** modify negative relationship dynamics; enhance stability, support from inside and outside the family, and a climate of hope, positivity, and closeness.
4. **Parenting skills:** learn the specific concepts, skills, and attitudes of Corrective Attachment Parenting that are effective with the attachment-disordered child (angry, oppositional, mistrustful, controlling, deceitful).
5. **Parents:** address historical and/or current issues that are unresolved and prevent effective functioning, including family-of-origin (prior loss, trauma, attachment difficulties) and current marital/relationship problems.

Corrective Attachment Therapy

The basic assumption is that the child's anger, defiance, aggression, and antisocial behavior is rooted in the early experience of compromised and disordered attachment. Therefore, the treatment framework must provide the physical, emotional, and interpersonal characteristics of secure attachment. The ingredients found in parent–child relationships leading to secure attachment are also made available in the therapist–child relationship, which requires the following:

- **Structure:** The therapist provides limits, rules, and boundaries similar to the clear and consistent structure provided by the sensitive and appropriately responsive caregiver. The structure is consistent yet flexible, changing in accordance with the child's development. For example, the therapist informs the child of the rules of therapy, and together they establish an explicit contract which defines their responsibilities and goals.
- **Attunement:** The therapist is "in synch" with the child's needs, emotions, and internal working model and provides the message: "I know what you need in order to feel safe, and I will offer it." For example, it is understood that the child's hostile and controlling demeanor is a defensive strategy, a reaction to feelings of vulnerability and fear.
- **Empathy:** Just as the healthy parent cares deeply about his or her child, the therapist conveys a heartfelt level of caring and compassion. The therapist is proactive and empathic and does not react negatively to hostility and distancing behavior. The message conveyed is: "How sad that those terrible things happened to you; I'm sorry you were treated badly; I understand what you feel and how much pain you must be in."
- **Positive affect:** Parents who foster secure attachment experience and display positive emotions with their children. The therapist also maintains a positive demeanor, particularly when the child is acting-out. This prevents the reenactment of dysfunctional patterns, such as mutual rejection or distancing. The message to the child is: "I will not allow you to control our relationship in unhealthy and destructive ways." This provides modeling of positive affect and appropriate boundary setting.
- **Support:** Parents of securely attached children provide a scaffold of support, a framework that supports the child as development unfolds. The therapist also provides support tailored to the developmental needs and capabilities of the child. Initially, the emphasis is on rules, expectations, and natural consequences. As therapy progresses, the focus shifts to reinforcing the child's independent achievements.
- **Reciprocity:** A positive reciprocal relationship involves mutual influence and regulation. The securely attached child achieves a "goal-corrected partnership" with caregivers, characterized by a sharing of control, values, feelings, and goals. The therapist guides the child toward a reciprocal relationship based on mutual respect and sensitivity. This begins with the establishment of a foundation for secure attachment (safety, protection, basic compliance); next the child learns to balance his or her own needs with those others.
- **Love:** Secure attachment is synonymous with love; the ability to feel a deep and genuine caring for and commitment to another. Children with

attachment disorder are generally incapable of experiencing and demonstrating love; they lack the early attachment relationship necessary to create that feeling. Therapy provides this relationship context and in doing so, guides the child to a place where love can be experienced. The open expression of loving feelings occurs with parents holding their children "in arms," eye-to-eye, face-to-face. Children, however, will only feel safe in experiencing and expressing love if the parent(s) are available to receive that level of intimacy and affection. Thus, therapy also helps the parents become emotionally available.

Treatment Strategies and Methodologies

The Holding Nurturing Process (HNP), in which the therapist holds the child eye-to-eye, face-to-face, is an "in arms" experience that promotes secure attachment. It stimulates infant and parent attachment behaviors practiced by most cultures throughout the world. It also reduces the effects of the alarm reaction caused by maltreatment, promotes self-regulation, and provides the structure necessary to meet the fundamental limit-setting needs of oppositional, angry, acting-out children. Specific prosocial coping skills (e.g., anger management) are learned within this therapeutic context.

Children with severe attachment disorder are extremely resistant to therapy and therapeutic relationships. There are a variety of therapeutic strategies that are effective in managing their resistance. The therapist is proactive, remains emotionally neutral, and avoids control battles. The therapist also conveys commitment and perseverance, provides paradoxical interventions, acknowledges underlying emotions, and holds the child accountable for choices and consequences.

Contracting with the child and parents is crucial to treatment success. Contracting increases commitment and motivation to change and provides clear structure, expectations, and goals. Therapeutic contracts are relationship agreements that establish a collaborative framework and facilitate accountability and opportunities for success. For example, we contract with the child to verbally express anger and defiance rather than act-out in physically and/or emotionally abusive ways.

Therapeutic goals include developing new belief systems, effectively dealing with emotions, learning prosocial coping skills, creating mastery over prior trauma, enhancing self-regulation, developing a positive sense of self, and facilitating healthy family dynamics. A variety of specific interventions are employed in order to achieve treatment goals.

1. *First-year attachment cycle:* this is an explanation of the attachment cycle that occurs in the first year of life. This intervention focuses on the correlation between need fulfillment, basic trust, and the establishment of secure attachment.

2. *Child's self-report and list:* the therapist and child develop a list of problems based on the information the child provides. This enables the therapist to understand the child's perceptions and forms the basis for a therapeutic contract regarding goals and changes.

3. *Rules of therapy:* the child is told the rules of therapy, which provides structure, a feeling of safety and security, and expectations for specific behavior. An example of a therapeutic rule is: "eye contact is required when you are speaking to the therapist."

4. *Review of historical information:* it is therapeutic to review with the child the relevant documents, such as social history, social service records, life books, and photographs. This reduces avoidance and denial regarding the child's traumatic history, creates a positive rapport, and facilitates revision of misinterpretation of prior events.

5. *Inner-child metaphor:* the child is gently guided back to an earlier time in life and encouraged to visualize him or herself as younger. This promotes understanding of early life experiences and provides a vehicle for the healing of attachment trauma.

6. *Psychodramatic reenactment:* treatment team members role-play scenarios from the child's life. This enhances genuine involvement in the therapeutic process, encourages honest expression of emotion and perception, and promotes emotional resolution and mastery.

Children with attachment disorder have internalized antisocial values, belief system, and patterns of relating: dishonesty, coercion, aggression, mistrust, betrayal, selfishness. Treatment must emphasize prosocial coping skills so that they can function successfully in families and in society. These children lack the ability to identify and manage emotions, communicate honestly, regulate impulses, and solve problems effectively. Teaching prosocial coping skills not only reduces acting-out, but also builds self-confidence and self-esteem. The child receives positive feedback from others (parents, siblings, friends, teachers), which reinforces positive behavior and enhances self-esteem. These skills include anger management, communication, and problem-solving.

CONCLUSIONS AND RECOMMENDATIONS

What are the solutions to the vast problems of violence in children and attachment disorder in families, the child welfare system, and society? Solutions are found in four areas: (1) attachment-focused assessment and diagnosis, (2) specialized training and education for caregivers (Corrective Attachment Parenting), (3) treatment for children and caregiver which facilitates secure attachment (Corrective Attachment Therapy), and (4) early intervention and prevention programs for high-risk families (Levy & Orlans, 1998).

Attachment disorder is one of the most easily diagnosed and yet commonly misunderstood parent–child disorders. Many social service and mental health professionals, although adept at assessing behavioral and emotional disorders in children, are not familiar with attachment concepts. Parents and other caregivers (e.g., foster parents) assume the responsibility of childrearing with challenging attachment-disordered children, often without the necessary information, training, and support. Adoptive and foster parents commonly feel frustrated, angry at the child and the "system," demoralized, disillusioned, and burned out. Specialized parenting skills are required in order to be successful in their parenting role. A significant amount of evidence accumulated over the past 25 years indicates that early intervention and prevention programs are effective for at-risk children and families (Guralnick, 1997; Ramey & Ramey, 1998). Early intervention and prevention programs have been shown to enhance parent–children attachment, foster children's cognitive and social development, and reduce later violence.

Forty million children will be entering adolescence in the next few years. A high percentage of these children were reared in physical and emotional environments that cultivated fear, rage, and violence. We, as a society, must not only address the needs of children damaged by attachment disorder, but also help educate professionals and parents so that future generations can engender secure attachment.

REFERENCES

Ainsworth, M. D. S. (1973). The development of infant-mother attachment. In B. M. Caldwell & H. N. Ricciuti (Eds.), *Review of child development research* (Vol. 3, pp. 1–94). Chicago: University of Chicago Press.

American Psychiatric Association. (1994). *Diagnostic and statistical manual of mental disorders* (4th ed.). Washington, DC: Author.

Barkley, R. A. (1990). *Attention deficit hyperactivity disorder: A handbook for diagnosis and treatment.* New York: Guilford Press.

Bell, C. C., & Jenkins, E. J. (1993). Community violence and children on Chicago's southside. *Psychiatry, 56,* 46–54.

Berkowitz, L. (1994). Guns and youth. In L. Eron, J. H. Gentry, & P. Schlegel (Eds.), *Reason to hope: A psychological perspective on violence and youth* (pp. 251–279). Washington, DC: American Psychological Association.

Berman, S. L., Kurtines, W. M., Silverman, W. K., & Serafini, L. T. (1996). The impact of exposure to crime and violence on urban youth. *American Journal of Orthopsychiatry, 66*(3), 329–336.

Bowlby, J. (1969). *Attachment and loss: Vol. 1. Attachment.* New York: Basic Books.

Butterfield, F. (1997, July 21). Juvenile courts on way out? *Denver Post.*

Cannon, A. (1997, August 10). Animal/human cruelty linked. *Denver Post.*

Children's Defense Fund. (1997). *The state of America's children: Yearbook 1997.* Washington, DC: Author.

Davis, J. A. (1998). Profile of a sexual predator. *The Forensic Examiner, 7*(1/2), 28–33.

DeMulder, E. K., & Radke-Yarrow, M. (1991). Attachment with affectively ill and well mothers: Concurrent behavioral correlates. *Development and Psychopathology, 3,* 227–242.

Dodge, K. A., Bates, J. E., & Pettit, G. S. (1990). Mechanisms in the cycle of violence. *Science, 250,* 1678–1683.

Douglas, J. (1998). Obsession: The hunters and the hunted. *The Forensic Examiner, 7*(5/6), 30–31.

Egeland, B., Pianta, R., & O'Brien, M. A. (1993). Maternal intrusiveness in infancy and child maladaptation in early school years. *Development and Psychopathology, 5,* 359–370.

Emery, R. E., & Laumann-Billings, L. (1998). An overview of the nature, causes, and consequences of abusive family relationships. *American Psychologist, 53*(2), 121–135.

Erickson, M., Sroufe, L. A., & Egeland, B. (1985). The relationship between quality of attachment and behavior problems in preschool and a high risk-sample. *Monographs of the Society for Research in Child Development, 50* (1–2, Serial No. 209), 147–166.

Eron, L. D., Gentry, J. H., & Schlegel, P. (Eds.). (1994). *Reason to hope: A psychological perspective on violence and youth.* Washington, DC: American Psychological Association.

Estrin, R. (1996, January 20). Juvenile firesetting. *Denver Post.*

Fromm, E. (1973). *The anatomy of human destructiveness.* New York: Holt, Rinehart & Winston.

Gallup, G. H., Moore, D. W., & Schussel, R. (1995). *Disciplining children in America: A Gallup Poll Report.* Princeton, NJ: The Gallup Organization.

Greenberg, M. T., DeKlyen, M., Speltz, M. L., & Endriga, M. C. (1997). The role of attachment processes in externalizing psychopathology in young children. In L. Atkinson & K. J. Zucker (Eds.), *Attachment and psychopathology* (pp. 196–222). New York: Guilford Press.

Guralnick, M. J. (1997). *The effectiveness of early intervention.* Baltimore, MD: Brookes.

Hann, D. M., Castino, R. J., Jarosinski, J., & Britton, H. (1991). *Relating mother-toddler negotiation patterns to infant attachment and maternal depression with an adolescent mother sample.* Symposium conducted at the biennial meeting of the Society for Research in Child Development, Seattle, WA.

Hare, R. D. (1993). *Without conscience: The disturbing world of psychopaths among us.* New York: Pocket Books.

Hofer, M. A. (1995). Hidden regulators: Implications for a new understanding of attachment, separation, and loss. In S. Goldberg, R. Muir, & J. Kerr (Eds.), *Attachment theory: Social, developmental, and clinical perspectives.* Hillsdale, NJ: Analytic Press.

Huesmann, L. R., Eron, L. D., Lefkowitz, M. M., & Walder, L. O. (1984). The stability of aggression over time and generations. *Developmental Psychology, 20,* 1120–1134.

Huston, A. C., Donnerstein, E., Fairchild, H., Feshbach, N. D., Katz, P. A., Murray, J. P., Rubinstein, E. A., Wilcox, B. L., & Zuckerman, D. (1992). *Big world, small screen: The role of television in American society.* Lincoln: University of Nebraska Press.

Jacobson, J. L., & Wille, D. E. (1986). The influence of attachment pattern on developmental changes in peer interaction from the toddler to the preschool period. *Child Development, 57,* 338–347.

Kachur, S. P., Stennies, G. M., Powell, K. E., & Lowery, R. (1994). School-associated violent deaths in the United States, 1992–1994. *JAMA, Journal of American Medical Association, 275,* 1729–1734.

Karr-Morse, R., & Wiley, M. S. (1997). *Ghosts from the nursery: Tracing the roots of violence.* New York: Atlantic Monthly Press.

Kolko, D. J. (1996). Child physical abuse. In J. Briere, L. Berliner, J. A. Buckely, C. Jenny, & T. Reid (Eds.), *APSAC handbook on child maltreatment* (pp. 21–50). Thousand Oaks, CA: Sage.

Levine, I. S. (1996). Preventing violence among youth. *American Journal of Orthopsychiatry, 66*(3), 320–322.

Levy, T., & Orlans, M. (1995). Intensive short-term therapy with attachment disordered children. In L. VandeCreek, S. Knapp, & T. L. Jackson (Eds.), *Innovations in clinical practice: A source book* (Vol. 14, pp. 227–251). Sarasota, FL: Professional Resource Press.

Levy, T., & Orlans, M. (1998). *Attachment, trauma and healing.* Washington, DC: Child Welfare League of America Press.

Lewin, T. (1998, May 22). Experts see a shift in the nature of school violence. *New York Times.*

Lewis, D. O. (1990). Neuropsychiatric and experiential correlates of violent juvenile delinquency. *Neuropsychology Review, 1*(2).

Lung, C. T., & Daro, D. (1996). *Current trends in child abuse reporting and fatalities.* Chicago: National Committee to Prevent Child Abuse.

Lyons-Ruth, K. (1996). Attachment relationships among children with aggressive behavior problems: The role of disorganized early attachment patterns. *Journal of Consulting and Clinical Psychology, 64*(1), 64–73.

Lyons-Ruth, K., Alpern, L., & Repacholi, B. (1993). Disorganized infant attachment classification and maternal psychosocial problems as predictors of hostile-aggressive behavior in the preschool classroom. *Child Development, 64,* 572–585.

Main, M., & Goldwyn, R. (1984). Predicting rejection of her infant from mother's representation of her own experience. *Child Abuse and Neglect, 8,* 203–217.

Main, M., & Hesse, E. (1990). Parent's unresolved traumatic experiences are related to infant disorganization attachment status: Is frightened and/or frightening parental behavior the linking mechanism? In. M. T. Greenberg, D. Chicchetti, & E. M. Cummings (Eds.), *Attachment in the preschool years: Theory, research, and intervention* (pp. 161–184). Chicago: University of Chicago Press.

Main, M., Kaplan, N., & Cassidy, J. (1985). Security in infancy, childhood and adulthood: A move to the level of representation. *Monographs of the Society for Research in Child Development, 50* (1–2, Serial No. 209), 66–104.

Main, M., & Solomon, J. (1990). Procedures for identifying infants as disorganized/disoriented during the Ainsworth Strange Situation. In M. Greenberg, D. Chicchetti, & E. M. Cummings (Eds.), *Attachment in the preschool years: Theory, research, and intervention* (pp. 121–160). Chicago: University of Chicago Press.

Martinez, P., & Richters, J. E. (1993). The NIMH Community Violence Project: II. Children's distress symptoms associated with violence exposure. In D. Reiss, J. E. Richters, M. Radke-Yarrow & D. Scharff (Eds.), *Children and violence.* New York: Guilford Press.

Mercy, J. A., & Rosenberg, M. L. (1996). *Children, guns, and schools.* A paper prepared for The Center for the Study and Prevention of Violence, University of Colorado, Boulder.

Morris, D. (1994). *The human animal.* New York: Crown.

National Center on Child Abuse and Neglect (NCCAN). (1993). *A report on the maltreatment of children and disabilities.* Washington, DC: Department of Health and Human Services.

National Center on Child Abuse and Neglect (NCCAN). (1995). *National child abuse and neglect data systems. Third national incidence study of child maltreatment.* Washington, DC: Government Printing Office.

National Committee for Injury Prevention and Control. (1989). *Injury prevention: Meeting the challenge.* New York: Oxford University Press.

National Institute of Mental Health. (1982). *Television and behavior: Ten years of scientific progress and implications for the eighties: Vol. 1. Summary report.* Washington, DC: U.S. Government Printing Office.

National Research Council. (1993). *Understanding and preventing violence.* Washington, DC: National Academy Press.

Osofsky, J. (1994). *Introduction: Caring for infants and toddlers in violent environments.* Arlington, VA: ZERO TO THREE.

Osofsky, J. D. (1995). The effects of exposure to violence on young children. *American Psychologist, 50,* 782–788.

Perry, B. D. (1994). Neurobiological sequelae of childhood trauma. In M. Murberg (Ed.), *Catecholamine function in posttraumatic stress disorder* (pp. 233–255). Washington, DC: American Psychiatric Press.

Perry, B. D. (1995). Incubated in terror: Neurodevelopmental factors in the 'cycle of violence.' In J. D. Osofsky (Ed.), *Children, youth, and violence* (pp. 45–63). New York: Guilford Press.

Raine, A. (1993). *The psychopathology of crime.* New York: Academic Press.

Ramey, C. T., & Ramey, S. L. (1998). Early intervention and early experience. *American Psychologist, 53*(2), 109–120.

Renken, B., Egeland, B., Marvinney, D., Mangelsdoft, S., & Sroufe, L. A. (1989). Early childhood antecedents of aggression and passive-withdrawal in early elementary school. *Journal of Personality, 57,* 257–281.

Richters, J. E., & Martinez, P. (1993). The NIMH community violence project: Children as victims of and witnesses to violence. *Psychiatry, 56*(1), 7–21.

Robinson, J. L., & Glaves, L. (1996). Supporting emotion regulation and emotional availability through home visitation. *Bulletin of ZERO TO THREE, 17*(1), 31–35.

Rosenberg, M. S., & Rossman, B. B. R. (1990). The child witness to marital violence. In R. T. Ammerman & M. Herson (Eds.), *Treatment of family violence: A sourcebook* (pp. 183–210). New York: Wiley.

Schore, A. N. (1994). *Affect regulation and the origin of the self: The neurobiology of emotional development.* Hillsdale, NJ: Erlbaum.

Schore, A. N. (1996). The experience-dependent maturation of a regulatory system in the orbital prefontral cortex and the origin of developmental psychopathology. *Development and Psychopathology, 8,* 59–87.

Showalter, J. E. (1983). The use and abuse of pets. *Journal of the American Academy Child Psychiatry, 22,* 68–72.

Shure, M., & Spivack, G. (1988). Interpersonal cognitive problem solving. In R. G. Price, E. L. Cowen, R. P. Lorion, & J. Ramos-McKay (Eds.), *14 ounces of prevention: A casebook for practitioners* (pp. 69–82). Washington, DC: American Psychological Association.

Slaby, R. G., & Guerra, N. G. (1988). Cognitive mediators of aggression in adolescent offenders: 1. Assessment. *Developmental Psychology, 24,* 580–588.

Spieker, S. J., & Booth, C. (1988). Maternal antecedents of attachment quality. In J. Belsky & T. Nezworski (Eds.), *Clinical implications of attachment* (pp. 300–323). Hillsdale, NJ: Erlbaum.

Sroufe, L. A., Carlson, E., & Shulman, S. (1993). *The development of individuals in relationships: From infancy through adolescence.* Unpublished manuscript.

Sroufe, L. A., Egeland, B., & Kreutzer, T. (1990). The fate of early experience following developmental change. *Child Development, 61,* 1363–1373.

Strauss, M. A. (1993). Ordinary violence, child abuse and wife-beating—what do they have in common? In D. Finkelhor, R. J. Gelles, G. T. Hotelling, & Strauss (Eds.), *The dark side of families.* New York: Sage.

Tapia, F. (1971). Children who are cruel to animals. *Child Psychiatry and Human Development, 2,* 70–77.

Troy, M., & Sroufe, L. A. (1987). Victimization among preschoolers: The role of attachment relationship history. *Journal of the American Academy of Child and Adolescent Psychiatry, 26,* 166–172.

U.S. Department of Justice. (1995). *Juvenile offenders and victims: A national report.* Washington, DC: Office of Juvenile Justice and Delinquency Prevention.

van der Kolk, B. (1996). The complexity of adaptation to trauma. In B. van der Kolk, A. C. Macfarlane, & L. Weisaeth (Eds.), *Traumatic stress* (pp. 182–213). New York: Guilford Press.

van der Kolk, B., & Fisler, R. (1994). Childhood abuse and neglect and loss of self-regulation. *Bulletin of the Menninger Clinic, 58*(2), 145–168.

Van IJzendoorn, M. H. (1995). Adult attachment representations, parental responsiveness, and infant attachment: A meta-analysis on the predictive validity of the Adult Attachment Interview. *Psychological Bulletin, 117,* 387–403.

Waters, E., Wippman, J., & Sroufe, L. A. (1979). Attachment, positive affect, and competence in the peer group: Two studies in construct validation. *Child Development, 50,* 821–829.

Werner, E., & Smith, R. (1992). *Overcoming the odds: High-risk children from birth to adulthood.* Ithaca, NY: Cornell University.

Yochelson, S., & Samenow, S. E. (1993). *The criminal personality: A profile for change.* Northvale, NJ: Jason Aronson.

Zahn-Waxler, C., Radke-Yarrow, M., Wagner, E., & Chapman, M. (1992). Development of concern for others. *Developmental Psychology, 28,* 126–136.

Zeanah, C. H., & Scheering, M. (1996). Evaluation of posttraumatic symptomology in infants and young children exposed to violence. In J. Osofsky & E. Fenichel (Eds.), *Islands of safety* (pp. 9–14). Arlington, VA: ZERO TO THREE.

BIOGRAPHIES

Terry M. Levy

Dr. Levy, Ph.D., DABFE, DAPA, Director of Psychological Services, is a Licensed Clinical Psychologist in Colorado and Florida, and a Board Certified Forensic Examiner in Clinical and Family Psychology. He is a clinical member of the American, Colorado and Florida Psychological Associations, American and Colorado Association of Marriage and Family Therapy, American Family Therapy Academy, American College of Forensic Examiners, American Psychotherapy Association, and the National Register of Health Service Providers in Psychology. He was founder and Director of the Family Life Center (Florida) and the Miami Psychotherapy Institute, and co-founder and President of the Board of Directors of the Association for Treatment and Training in the Attachment of Children (ATTACh). Dr. Levy has been providing psychotherapy treatment and training for over 25 years. He has taught seminars on therapeutic issues for the American Psychological Association, the American Professional Society on the Abuse of Children, and numerous mental health, social service, and school systems nationwide. Currently, Dr. Levy has a private practice with Evergreen Consultants in Human Behavior. Dr. Levy is co-author of *Attachment, Trauma and Healing* (1998, Child Welfare League of America) and editor of *Handbook of Attachment Interventions.*

Michael Orlans

Michael Orlans, M.A., DABFE, DAPA, Director of Training, is in private practice and Director of Training with Evergreen Consultants in Human Behavior. He is the former clinical supervisor for the Domestic Intervention Program - State Attorney's Office, eleventh judicial circuit of Florida. He is a certified criminal justice specialist and is a Marriage and Family Therapist with over 25 years of clinical experience. In addition to his work in private practice and in the public sector, Michael has taught on the faculty of several universities. He is a nationally known lecturer and trainer, with expertise in working with severely emotionally disturbed children and their families. He has served as a consultant to therapeutic foster care programs, child welfare agencies, and is on the Advisory Council of the National Alliance for Rational Children's Policy. He is co-founder of the Association for Treatment and Training in the Attachment of Children. He is also a Diplomate, Board Certified Forensic Examiner, and founding executive board member and Diplomate of the American Psychotherapy Association. Co-author, *Attachment, Trauma and Healing,* Child Welfare League of America Press.

2

TREATING ADHD AS ATTACHMENT DEFICIT HYPERACTIVITY DISORDER

RANDALL D. LADNIER

Family Counseling Center of Sarasota County, Inc.
Sarasota, Florida

ALICE E. MASSANARI

InterActive Family Resources
Asheville, North Carolina

BUILDING A DEVELOPMENTAL MODEL FOR UNDERSTANDING ADHD

Introduction

Ashley was 8 years old when she was brought to our community mental health agency. Her mother had finally run out of patience with her compulsive, ritualistic behaviors. The little girl presented with a desperate need to control what she wore each day. She would only wear one "special" pair of underpants, which she used to masturbate whenever she was stressed. Each morning, the demands for Ashley to get dressed for school led to verbal and physical combat between mother and daughter.

Tony was 15 years old when he was ordered by the court to receive treatment at the same agency for having sexually molested a 6-year-old girl in the foster home where they had been living. He had been removed from his mother's care 2 years earlier because he had threatened her with physical violence and could not be controlled by his teachers at school. His aggressive history notwithstanding, he was an extremely shy child who avoided eye contact and did not make friends with peers.

Rick was 18 years old when he voluntarily sought counseling at this agency. He was drinking beer and smoking marijuana on a regular basis and was failing in his effort to obtain a diploma at an alternative high school. He had moved out of his mother's home and was living with his abusive girlfriend and her mother. An intelligent and gentle young man, he suffered with chronic, immobilizing anxiety and avoided interaction with male peers.

At first glance, Rick, Tony, and Ashley appeared to be quite dissimilar in terms of their presenting problems, other than a pattern of inappropriate, self-defeating behaviors. Looking at their psychosocial histories, however, we discovered a surprising commonality in their diagnoses. Each one appeared to meet diagnostic criteria for Attention Deficit Hyperactivity Disorder (ADHD), a mental disorder identified in children and adults and characterized by chronic problems with inattention, hyperactivity, and impulsivity (DSM-IV, American Psychiatric Association, 1994).

Rick had been labeled ADHD in elementary school and had been prescribed Ritalin, which he adamantly rejected after a brief trial. Tony had been diagnosed as having ADHD in the sixth grade and treated with Ritalin, which he eventually refused because of its unpleasant side effects. He recalled with pride: "That stuff made me sit down but it couldn't make me shut up!" Ashley had never been assigned an official diagnosis of ADHD. She was oppositional and defiant at home and quite hyperactive in the therapy room with her mother, but she behaved fairly well at school where she had been placed in a program for "gifted" children. In the classroom, she evidenced anxiety related to her fear of making mistakes in her work and expressed anger toward herself. On occasion, she had been admonished by her teacher for not paying attention in class and for her impulsive behaviors. Had Ashley been a male child, and more physically aggressive in the classroom, she might have been identified by her teacher as having ADHD and referred to a physician for medication.

Detailed biopsychosocial assessments of these clients also revealed a fascinating coincidence: in one way or another, each one had been overstimulated and undersoothed in the first 2 years of life. Behaviorally, each of them could be described, at times, as inattentive, impulsive, or hyperactive, depending on where he/she was, who he/she was with, and what he/she was doing. None of the three had learned to maintain friendships with peers. They shared significant deficits in two important areas of their personality: (1) the ability to regulate emotions and behaviors and (2) the ability to form healthy relationships with others.

Intrigued by the realization that three children who appeared to have so little in common could meet the diagnostic criteria for ADHD, we decided to look more closely at their family histories. We eventually learned that each of them had failed to form an adequate bond with their childhood caregivers, or with anyone else, for that matter. Instinctively, we turned to attachment theory for principles with which to interpret what we had found. We learned that the word "attachment" is generally used to describe the enduring, affectionate, reciprocal bond that develops between a healthy child and his responsive caregiver in the first 2 years of life. Consistent with what has recently been discovered about the organization and development of a child's brain, attachment theory is based on the assumption that a child's neurological and emotional development is largely determined by the quality of his interactions with primary caregivers during those early years. In other words, the nurturing interactions that enable caregiver and infant to form a secure attachment also determine the neural development of that child's brain.

We soon realized that the two main areas of deficits we had identified in our three young clients (self-regulation and relating skills) were consistent with the classic symptoms of an attachment-disordered child. Consider the commonalities in diagnostic criteria for the DSM-IV diagnosis of ADHD and the symptoms of attachment disorder: impulsivity, hyperactivity, and impaired social functioning (Cline, 1979).

This realization raised two obvious questions: (1) Is there a causal connection between attachment failure and ADHD? and (2) Would it be possible to create a developmental model, based on attachment theory, that would provide a valid and credible explanation for the origin of ADHD and suggest a treatment plan that could offer a child more than temporary relief from symptoms?

ADHD as a Psychiatric Diagnosis

Attention Deficit Hyperactivity Disorder has recently become one of the most controversial of the so-called psychiatric disorders because of a startling rise in its incidence in American children and because it is most commonly treated with the psychostimulant medication Ritalin, which is classified as a controlled substance. This escalating controversy has increased competition, criticism, and conflict between parents, teachers, physicians, and psychotherapists. This debate continues to swirl, primarily in the areas of education and mental health, because we have not been able to reach a consensus on the origin and nature of ADHD or on the most appropriate ways to treat the disorder.

Diagnostic Criteria

The diagnostic criteria for ADHD contained in the latest version of the *Diagnostic and Statistical Manual of Mental Disorders* (DSM-IV, American Psychiatric Association, 1994) are the result of a 25 year effort to create an accurate and reliable list of symptoms for the identification and treatment of a complex syndrome of cognitions, emotions, and behaviors first observed in hyperkinetic children (Barkley, 1998). They are derived from factor analysis of commonalities found in studies of parent and teacher ratings of children diagnosed as having ADHD and are divided into two separate clusters: (1) symptoms of inattention and (2) symptoms of impulsivity-hyperactivity (Lahey et al., 1994).

In order to qualify for a diagnosis of *ADHD, Predominately Inattentive Type,* or ADHD, *Predominantly Hyperactive-Impulsive Type,* a child must be identified as having at least six discrete symptoms from one of the two cluster areas—symptoms that represent clear evidence of significant impairment in social, academic, or occupational functioning. The symptoms must be present in at least two settings (e.g., at school and at home) and must have been present before the age of 7 (DSM-IV, American Psychiatric Association, 1994).

Suggested Etiologies

A review of the literature relating to ADHD includes a wide variety of theories or models to explain the origin of the disorder—such theories as genetic transmission, neurological anomalies, imbalance of brain chemicals, viral infections, prenatal and postnatal toxins, faulty childrearing, socioeconomic status, food additives, and lead poisoning. Although no one has proven a specific chromosomal connection with ADHD, the most popular explanation for the cause of ADHD is known as the biomedical or neurobiological model, which basically asserts that ADHD is a genetically inherited medical disease that results in early neurological impairment of the developing brain. While researchers continue to search for a better model with which to explain the origin of this complex problem, psychostimulant medications such as Ritalin, Dexedrine, and Adderall have become the treatment of choice for the symptoms of ADHD in children (Barkley, 1998).

Current Prevalence

By 1995, some respected professionals were becoming skeptical of the disease model for ADHD, not only with regard to its assumed etiology of genetic transmission, but even to its classification as a mental disorder and especially to the use of psychostimulant medications for treating children (Armstrong, 1995). Today, many advocates for children are gravely concerned by the rising numbers of elementary school students being diagnosed with ADHD and by the expanding array of psychotropic medications with which they are being treated. It has been suggested that the increase is more likely attributable to a broadening of the diagnostic criteria in DSM-IV, or to changes in the basic structure of families in America, than to some mysterious anomaly in the gene pool of this country (Merrow, 1995).

At a conference in 1996, Gene R. Haislip of the U.S. Drug Enforcement Administration reported that prescriptions of Ritalin for the treatment of ADHD had increased 500% since 1990, and prescriptions of amphetamines for treating the disorder had grown by 400%. He also noted that between

7 and 10% of male, school-age children in America were being prescribed these drugs at one time or another, with the percentage rising to 15% in some locations (Haislip, 1996).

It is possible to identify a variety of other factors that are contributing to the prevalence of the disease model for treatment of ADHD in America. Health insurance companies, for example, find it costs them less to pay a physician to treat the child pharmaceutically than for a psychotherapist to see the family for regular sessions. Physicians today generally assume that medication is the first and most effective treatment for a child diagnosed as having ADHD. Local school boards may prefer to see ADHD as a medical problem because they are able to obtain increased funding for educating children with disabilities. Teachers may suggest medical treatment for children who display symptoms of ADHD because they believe that medication will quickly eliminate disruptions in their classrooms. And many parents, feeling accused and frustrated, find it easier to accept the idea that their child has a permanent, genetic disease than the thought that they might have contributed to a problem that is temporary and treatable.

The Connection between Childhood Trauma and Insecure Attachment

The commonalities identified in the detailed histories we obtained from Ashley, Rick, and Tony gave us our first clues that there might be a causal relationship between childhood trauma, attachment difficulties, and the symptoms of ADHD. We immediately set out to learn as much as we could about bonding and attachment between caregiver and child.

Principles of Attachment Theory

The first premise of attachment theory is that nature provides a biologically based, species-specific system of attaching behaviors that serve to bring newborn animals and humans closer to caregivers for the sake of safety and survival (Ainsworth, Blehar, Walters, & Wall, 1978). Beyond this basic function, attachment with the caregiver enables a child to succeed in the accomplishment of other tasks essential for normal development (Levy & Orlans, 1998). The book entitled *Attachment, Trauma, and Healing* identifies additional benefits that are derived from healthy attachment between caregiver and child.

- Learning basic trust and reciprocity for use in future relationships
- Developing the capacity for self-regulation of affect and behavior
- Forming an identity that includes a healthy sense of self-worth and autonomy
- Establishing a set of moral values derived from empathy, compassion, and conscience

- Developing the resourcefulness and resilience required to withstand trauma and stress
- Experiencing the stimulating interactions required for development of a healthy brain

When a child experiences sufficient reciprocal, affectionate interactions with a caregiver who is available and responsive to the child's appropriate needs, the bond that is formed between them is described as "secure" attachment. Children who fail to receive consistent, predictable, reassuring responses from their primary caregivers are likely to learn patterns of attachment that are described as "insecure." Secure attachment develops as the result of thousands of daily interactions in which the caregiver achieves affective attunement with the child, eventually becoming a secure base from which the child can tackle new tasks and new relationships with the consistent expectation of being successful and loved. Insecure attachment develops within the context of daily interactions in which the child experiences the caregiver as unavailable, unresponsive, unreliable, intrusive, or coercive. The child who does not learn to use his or her caregiver as a secure base for exploring the world will be at high risk for pathological development in the areas of social relationships, emotional development, behavioral control, and cognitive capacity (Hughes, 1997).

Types of Insecure Attachment

Mary Ainsworth developed a method for identifying and classifying the attachment patterns of year-old infants in a laboratory experiment. In addition to the category of secure attachment, she identified two subtypes of insecure attachment: (1) Ambivalent and (2) Avoidant. Infants labeled Ambivalent were in general more likely to cling to their mother in an unfamiliar environment and less willing to explore on their own. When separated from her, they appeared anxious, agitated and tearful. When the mother returned, they tended to seek contact with her but simultaneously rejected her attempt to soothe them. Infants labeled Avoidant generally gave the impression of being independent and self-sufficient. They tended to explore the unfamiliar environment with little concern for their mother's whereabouts, just like the children who were considered to be securely attached. They differed, however, from the securely attached children in that they seemed unaffected by separation from the mother and either rejected or avoided her when she returned (Ainsworth & Wittig, 1969).

Numerous researchers have since demonstrated significant correlation between infants with patterns of avoidant attachment and mothers who suppress anger, lack tenderness in holding and touch, and who are intrusive and rejecting in their interactions with the child (Belsky, Rovine, & Taylor, 1984; Grossmann, Grossmann, Spangler, Suess, & Unzner 1985; Lyons-Ruth, Connell, Zoll, & Stahl, 1987).

In 1990, researchers identified a third pattern of insecure attachment in some year-old infants. They have labeled it as Disorganized Attachment, using this term to describe infants who apparently lack a consistent strategy for organizing their comfort-seeking behaviors in times of stressful separation or reunion with the mother. Their disorganized reactions include such things as apprehension, helplessness, and depression. Some demonstrate desperate reactions like prolonged motor freezing or dissociation alternated with agitation in unpredictable ways (Main & Solomon, 1990).

It has subsequently been suggested that the incidence of infants who display patterns of insecure, disorganized attachment increases significantly in families where the mother is an adolescent, abuses alcohol, or suffers from depression (Lyons-Ruth, Repacholi, McLeod, & Silva, 1991). Other studies have documented a relationship between disorganized patterns of attachment in infancy and aggressive, controlling behaviors in childhood (Lyons-Ruth, Alpern, & Repacholi, 1993; Wartner, Grossmann, Fremmer-Bombick, Suess, 1994).

Those of us who are interested in developing a model to explain the origin of ADHD should pay close attention to the research that has described the characteristics of insecure attachment patterns in children and adolescents. There are obvious similarities between the behaviors of the Avoidant child and the symptoms of ADHD. Descriptions of the child with a disorganized pattern of attachment reminds us of children diagnosed as having both ADHD and Conduct Disorder (DSM-IV, American Psychiatric Association, 1994).

New Information about the Development of a Child's Brain

In recent years, brain-imaging technologies have enabled researchers to gather new information about the development of a child's brain. Whereas experts once believed that a child's brain is basically inherited and its development determined by his or her unique set of genes, we have recently learned that the individual brain is built from a complex mixture of heredity and experience. Heredity provides some basic parameters for brain potential, but experience determines the quality and quantity of the circuits that are formed and the way in which the brain is organized to process information. The quantity and quality of stimulating interactions between caregiver and child, in the first 3 years of his life, play the largest roles in determining his emotional development, learning potential, and level of adult functioning. Children who enjoy consistent, nurturing interactions with caregivers in their early years become more resilient, emotionally and biologically, and are better able to tolerate stress in later years (Karen, 1994).

The newborn brain is wired only for controlling those functions that are essential for survival; respiration, heart rate, and reflex movements. At birth, the infant has approximately 100 billion brain cells which are not yet connected to form the networks that will allow thinking and learning to occur. By the age of 3 the baby will have formed about 1000 trillion of these neural connections—about twice the number found in the adult brain. Around the age of 11 a child's brain performs a sort of self-pruning, ridding itself of excess neural connections and retaining the strongest ones—those that have been used most often. Connections that have been repeated most consistently in childhood are most likely to become permanent. Conversely, a lack of experience may create deficits in neural connections. For example, the child who is rarely spoken to may have insufficient neurons for mastering language, and the child who is rarely played with may not have enough circuits for healthy social attachment. The best insurance against emotional and behavioral problems in childhood is the formation of healthy attachment between a child and a consistently available, sensitively responsive, nurturing caregiver (Begley, 1997; Perry, 1994).

The Effects of Childhood Trauma on the Developing Brain

By the mid-1980s, researchers had formulated a fair understanding of the impact of extreme trauma on combat veterans of the Vietnam War and on other traumatized adults. In her study of traumatized children, Dr. Lenore Terr identified patterns of symptom clusters that resembled other psychiatric disorders, such as conduct disorders, anxiety disorders, and Attention Deficit Hyperactivity Disorder (Terr, 1991).

Within the past 5 years, a growing body of research has increased our understanding of the ways in which trauma affects the neurological and neurobehavioral functioning of children. We have learned that sustained traumatic experiences like childhood abuse and neglect, or failure to form a secure attachment in the early years of life, can create a chronic state of hyperarousal in a child and alter the neuroendocrine activities of his brain, causing him to become trapped in the "fight-or-flight-or-freeze" response (Perry, 1994; van der Kolk, 1994).

When a child's stress-response system remains activated for an extended period of time, his brain will adapt in ways that enable him to live with the perception of constant threat. Changes in the neurology of his brain result in cognitive, emotional, and behavioral changes. The following list presents neurobehavioral problems likely to be identified in traumatized, hyperaroused children (adapted from Perry, 1995). It is important to note that the list includes three major symptoms of ADHD: hyperactivity, impulsivity, and impaired social functioning.

- Hypersensitivity and overreaction to neutral stimuli
- Autonomic hyperarousal and motor hyperactivity
- Increased startle response
- Profound sleep disturbance
- Problems controlling emotions

- Cognitive deficits and distortions
- Impaired social functioning
- Impulsivity and aggression

An Etiologic Model for ADHD Based on Developmental Trauma

After 4 years of working with and learning about children and adults diagnosed with ADHD, we have assembled a developmental model that explains the origin of this controversial disorder. It is based on the belief that attachment trauma in early childhood results in developmental deficits which, in the absence of remedial parenting, are likely to be manifested as the symptoms of ADHD. Our model can be stated very simply:

Bonding breaks → Attachment deficits → Symptoms of ADHD.

This model is based on information gleaned from detailed assessments of all our clients diagnosed with ADHD, each of which provided clear evidence of what we have come to call a "bonding break"—an event or combination of events, occurring prenatally or postnatally, that causes physiological trauma and developmental arrest and interferes with a child's opportunity to form a secure attachment with a caregiver. This developmental model answers the question: "What causes ADHD?" It also provides a treatment plan that makes medication the treatment-of-last-resort and emphasizes the reparation of attachment deficits instead of attention deficits.

Ours is a model based on theory, but is consistent with the scientific research presented above. It begins with three major assumptions: (1) a child who meets diagnostic criteria for ADHD has experienced some sort of bonding break(s) before the age of 2, (2) the bonding break(s) have interfered with the process of healthy attachment between caregiver and child and created developmental deficits in the child, and (3) the family system in which the child grew up was not healthy enough to overcome those deficits. These assumptions have been derived from the histories of nearly 50 clients diagnosed as having ADHD. Every one that we studied over the past 4 years provided real evidence of at least one significant bonding break. We theorize that the failure to attach to an adult caregiver constitutes psychological and physiological trauma that interferes with an infant's neurological and hormonal maturation, resulting in developmental delays (attachment deficits) that are reflected in emotional and behavioral problems. Hence the simple formula

Bonding breaks → Attachment deficits → Symptoms of ADHD.

Figure 1 illustrates the dynamics of this model.

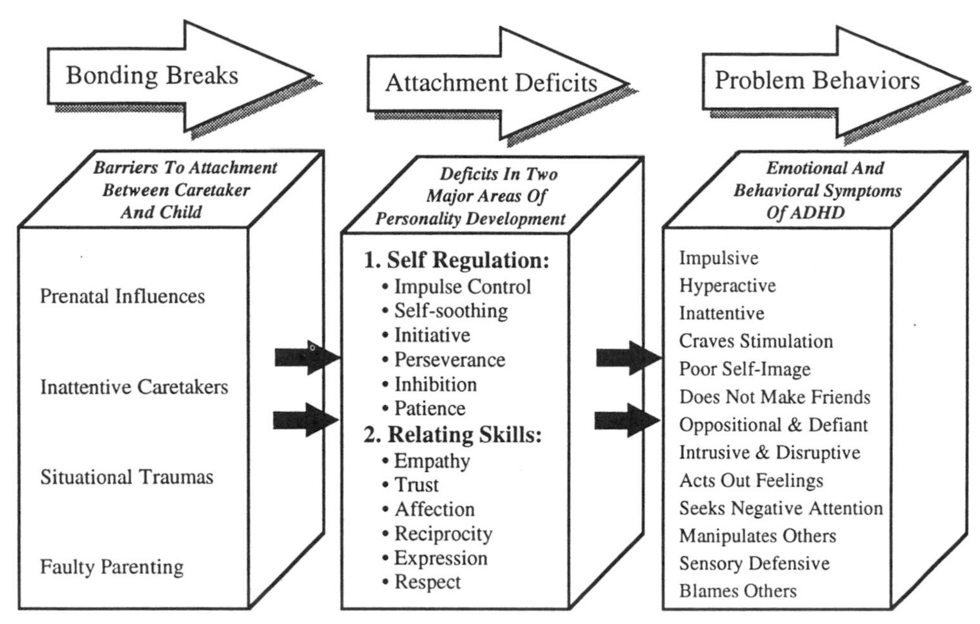

FIGURE 1. A developmental model for the origin of ADHD.

Four Basic Types of Bonding Breaks

From our collection of detailed biopsychosocial histories of children and adults diagnosed with ADHD, we have identified four relatively discrete categories of bonding breaks: Prenatal Influences, Inattentive Caregivers, Situational Traumas, and Faulty Parenting. The subjects we studied were more likely to have experienced a combination of bonding breaks, sequentially or simultaneously, involving more than one category, than to have experienced a single break.

Prenatal Influences

A healthy newborn arrives in this world already programmed, like a heat-seeking missile, to attach to a suitable caregiver. Some newborns, however, arrive in a state of distress and extreme hyperarousal. These infants are not programmed to seek out a caregiver because their state of emotional alarm prevents them from responding to attaching cues in their caregivers. Because prenatal influences that have interfered with nature's programming were so often identified as the culprits responsible for attachment problems in our clients, we have concluded that the relevance of this category of bonding break has not been fully appreciated by therapists and developmental researchers. In fact, it is rarely mentioned as a possible contributing factor in the development of ADHD.

For example, researchers suggest that mothers who smoke cigarettes during their pregnancy are three times more likely to give birth to a child who will be diagnosed as having ADHD (Milberger, Biederman, Faraone, Chen, & Jones, 1996). This finding is consistent with our hypothesis that prenatal exposure to certain chemical compounds or toxins may cause hyperactivity and impulsivity in a child. In our clinical practice, we also identified subjects who had undoubtedly been exposed prenatally to stress hormones such as cortisol, adrenalin, and norepinephrine by mothers who experienced chronic anxiety or panic attacks during pregnancy. Cortisol is apparently a major threat to the healthy development of a child's brain. Produced by the adrenal gland, cortisol helps the human body respond in a time of crisis by moderating its level of stress-response, but elevated levels of this hormone interfere with the building of neural pathways and may even have the effect of dissolving established connections (Begley, 1997; Perry, 1994).

Excess levels of cortisol are also believed to damage the brain's hippocampus, thereby interfering with memory function and lowering an individual's ability to control his/her emotions. Cortisol and other stress hormones also have the potential for inhibiting areas of the brain that regulate attention (Nachmias, Gunnar, Mangelsdorf, Parritz, & Buss, 1996).

We identified subjects who were born to mothers overwhelmed by stress due to being depressed, abused, or frightened. We found others who had been severely traumatized during labor and delivery. We documented numerous examples of mothers who failed to bond with the child they were expecting because it was the wrong sex, had the wrong father, or was arriving at the wrong time. Surprisingly, several mothers confessed that they had felt extremely depressed when they learned they were expecting a male child.

We discovered an intersting example of a prenatal bonding break in the case of fraternal twins, one of whom met diagnostic criteria for ADHD. Tim, 7 years old, had become a serious behavior problem at home and at school, in spite of taking his Ritalin as prescribed. His brother was compliant and more social at school and behaved fairly well at home. The twin with ADHD had been born with such an elevated level of anxiety that he could not be easily calmed or soothed. As an infant, he was irritable, agitated, and hyperactive. He had resisted swaddling by arching his back and slapping his hands together over his head. He seemed to lack interest in closeness and nurturing and had failed to make consistent eye contact with his caregivers. One day, in family therapy, his mother mentioned that her obstetrician had delivered the twins by Caesarean section because she had been carrying Tim in an awkward position, with his brother lying heavily on top of him. It had not occurred to her that the pain her son experienced *in utero,* and the hyperarousal he exhibited postnatally, had interfered with her opportunity to form a secure attachment with him.

We learned that Ashley, our 8-year-old with the underpants obsession, had apparently been exposed to chronic and acute levels of stress hormones before her birth. Ashley's mother acknowledged that her pregnancy had been very stressful. About the same time she conceived her daughter, she had become bored with her marriage and had been involved sexually with another man. She began to worry that she would give birth to a racially mixed child, thereby revealing her infidelity. In addition, her husband's family had placed great emphasis on the need for a male child to carry on the family name. She began to feel like a failure when she learned she would have a daughter. These two factors contributed to her frequent panic attacks during the last months of her pregnancy. Ashley's parents divorced before she was 3 years old. In therapy, her mother admitted that she had never felt a close bond with Ashley but stated she did feel close to Ashley's 6-year-old sibling, a male.

Inattentive Caregivers

A second type of bonding break occurs when a healthy infant is born to caregivers who fail to provide the minimum amount of warmth required for the heat-seeking missile to find its target. Such caregivers neglect their infants because they are self-centered and lack empathy for others or because they lack sufficient information or motivation to provide the nurturing and soothing interactions an infant needs. Interactions that contribute to attunement, synchronicity, and attaching responses in the infant include eye contact, soothing words and touch, breast-feeding, holding, rocking, smiling, and singing. It is the quantity and quality of these interactions that determine a child's neurological development as well as its capacity for empathy, trust, self-soothing, impulse control, initiative, and conscience.

One mother of an 8-year-old boy who was having severe behavioral problems at school first told us that she had not encountered any problems with him during pregnancy or in toddlerhood. She said that she had loved him from the first minute she saw him and that they had bonded perfectly with the exception of one small problem. When asked to explain what had happened, she exclaimed: "I just couldn't stand him pulling at me! It was disgusting! I was so grossed out, I didn't know what to do! My husband came home and found me crying my eyes out and he said: 'Hey, if you don't want to nurse him, don't!' So I stopped nursing him and everything was fine after that." This appears to be an example of a mother who lacked empathy for her nursing infant.

An inattentive caregiver, disappointed and angry about the sex of an infant, might ignore or reject that child. An inattentive mother may not have the patience to teach a child to nurse successfully or the motivation to get up and feed him during the night. A self-centered father might be jealous of the attention being given to the infant by his mother and withhold his cooperation or sabotage the child's care in some way. An inattentive caregiver might lose patience with a crying infant and handle him in a rough or abusive manner. A self-centered parent might reject a child who is born with a major birth defect.

In a therapist's office, the inattentive caregiver might say something like: "What did I do to deserve a child like this?" or "This child never thinks about what I need!" or "This child lives to make my life miserable!" Whatever the specifics of this caregiver's problems, such a person probably did not receive sufficient attention and empathy from his/her own caregivers in childhood. Herein lies a cogent explanation for the intergenerational transmission of ADHD. Parents who have received too little attention and empathy in their childhood may have a difficult time providing these emotional supplies to their own infants.

Rick, our young high school student, was born to an inattentive mother. As an adolescent, she had been required to serve as primary caregiver to her younger siblings. As an adult, she failed to give her own children the attention, affection, and nurturing they needed because she was busy getting her own needs met by the men in her life. When Rick was 15, she kept him out of school for a year so that he could serve as the full-time babysitter for his half-sister and half-brother, ages 1 and 3.

Situational Traumas

A third category of bonding breaks includes a variety of conditions and events that occur outside the control of the primary caregivers. These situations disrupt or preclude the formation of secure attachment between primary caregiver and infant. For example, an infant who is born prematurely might be deprived of consistent physical contact with his primary caregiver for weeks or months while being confined to a hospital. This explains the strong correlation between premature birth and ADHD (Szatmari, Saigal, & Rosenbaum, 1990).

There must be an unlimited number of ways in which a child can experience a situational break in bonding. The incidence of ADHD in children who have been adopted or placed in foster care is apparently quite high. Many infants are separated from a primary caregiver by illness or death. Some infants may be born with unrelieved postpartum pain from an injury or from frequent cholic or earache. Some are born with physical defects or with disabilities in sight or hearing that impede the process of attaching with caregivers. Primary caregivers may have disabilities, mental or physical, that prevent them from performing those activities that are likely to elicit attaching behaviors from their infant.

Tony, our physically and sexually aggressive teenager, was born to a mother who was impaired in speech and hearing. Her intellectual functioning was well below average. She abandoned Tony when he was 15 months old, leaving him in the care of elderly grandparents. When she returned 4 years later, he refused to accept any affection from her and was physically violent toward her.

Faulty Parenting

This last category of bonding breaks is prevalent in home situations where it is very difficult, if not impossible, for a child to overcome the harmful effects of early trauma. Typically, our clients diagnosed as having ADHD have grown up in families that shared three debilitating characteristics: (1) the absence of a healthy relationship between two caring adults; (2) a pattern of exposure to yelling, criticism, sarcasm, and violence; and (3) parenting that lacked respect, discipline, structure, and consistency.

Clearly, this raises an interesting question. Does the experience of growing up with inadequate parents constitute a bonding break in itself or is it coincidence that all of our ADHD clients received inadequate parenting? After all, it is just as easy to find examples of children who experienced an early bonding break but did not develop the symptoms of ADHD as it is to find children who grew up with faulty parenting and turned out to be fairly normal. With our clients, we found many examples of someone having experienced bonding breaks in all four categories. Faulty parenting was the only category of bonding break that showed up in the biopsychosocial assessment of every client who met diagnostic criteria for ADHD.

Some children apparently experience a bonding break around the age of 15 to 18 months. This is normally the time when caregivers begin to say "no!" and to set limits on a child's behavior. When corrected and disciplined by parents who provide sufficient empathy and reassurance, the child will learn that a caregiver can disapprove of his behavior while remaining emotionally supportive and available. When the discipline is coercive, abusive, or rejecting, the child might feel shamed and threatened and withdraw from caregivers, thereby breaking the incipient psychological bond (Hughes, 1997).

What we seem to have discovered is that children who are deprived of the opportunity to witness caring, respectful, and trusting relationships; who fail to receive adequate discipline, structure, and boundaries from their caregivers; and who are exposed to chronic conflict and chaos are unlikely to overcome early bonding breaks and attain healthy development. That is to say, children who experience early attachment trauma and are raised with "good enough" parenting will probably avoid the label of ADHD, while others (especially male children), who encounter similar breaks but grow up in a family where the parenting is faulty, are likely to be diagnosed as having the disorder and treated with psychostimulant medication by age 7.

Attachment Deficits

We have coined the phrase "attachment deficits" to designate specific shortcomings in the emotional development of a child. They refer to character traits that appear to be absent or underdeveloped in a child, as evidenced by emotions and behaviors that set him apart from his peers. These deficits might correspond to specific regions of neural circuitry in a child's brain that have not developed normally. Science tells us that developing circuits in the various sectors of the human brain mature on different schedules. Motor skills, for example, begin to develop prenatally. Math and logic skills do not begin to be wired in a child's brain until about the age of 15 months. Two important spheres of neural circuitry that appear highly vulnerable to bonding breaks are the areas of the brain that determine emotional control and social attachment. These two areas of the prefrontal cortex begin to develop by the time an infant is 6 months old. A child who does not receive proper stimulation and consistent nurturing interactions from caregivers in the first year of life is unlikely to develop sufficient neural circuitry required for emotional control and social attachment (Lach, 1997).

It is not surprising that researchers, using magnetic resonance imaging scans, recently found slight abnormalities in the size of the right prefrontal-striatal systems in some male children diagnosed as having ADHD (Castellanos et al., 1996). After all, this is the area of the brain that is responsible for the management of emotional control and social attachment. In our model for explaining the origin of ADHD, we refer to these areas as *self-regulation* and *relating skills.* From our clinical experience, nearly all children diagnosed with ADHD appear to exhibit deficits in these two areas of personality.

Deficits in Self-Regulation

This category of attachment deficits includes all those human qualities that would ordinarily flow from a healthy individual's ability to regulate his own thoughts, feelings, and behaviors. Although we are only identifying six of them here, their total possible number cannot be estimated. Taken collectively, these capacities represent a major part of what we have come to call a "self." When they are lacking in a child, he will be at high risk for a diagnosis of ADHD.

• ***Impulse control:*** A child learns to exercise control over his feelings and behaviors when he has received sufficient limits, guidance, and discipline from adult caregivers. A child with attachment deficits lacks internalized limits. Unable to distinguish between thoughts and actions, he or she simply reacts to what he or she is feeling or wants. Children who lack impulse control will usually be perceived as hyperactive.

- ***Self-soothing:*** A child learns healthy ways to soothe his own anxiety when he or she has received adequate and consistent soothing from an adult caregiver. When that is lacking, the child learns to substitute stimulation for soothing and lives in a chronic state of hyperarousal.
- ***Initiative:*** A child learns autonomy and self-motivation when he or she has achieved self-confidence and self-discipline and has developed internal values and goals with the help of an adult caregiver. A child who is insecurely attached has not yet developed a permanent sense of self and is unable to activate internal motivation.
- ***Perseverance:*** A child learns to make the sustained effort required for success when he or she is encouraged, validated, and given realistic expectations. A discouraged child usually receives a steady dose of negative feedback in place of the encouragement and validation needed to persevere. He or she eventually begins to see him- or herself as incompetent and helpless and is easily distracted.
- ***Patience:*** A child learns to wait patiently for what he or she wants when taught to accept delayed gratification. Children who fail to experience sufficient limit-setting and restraint from caregivers will generally act impulsively and will be perceived as intrusive or coercive at home and in school.
- ***Inhibition:*** A child learns healthy inhibitions that help him or her to avoid danger and self-injury when he or she receives consistent guidance and protection from his or her caregivers. When he or she is neglected or ignored he fails to develop proper awareness or respect for potential dangers and is unable to anticipate or avoid injury. Instead, he or she frequently engages in high-risk behaviors and is often viewed as being fearless.

Deficits in Relating Skills

This category of attachment deficits includes those qualities which, when present in healthy human beings, enable them to form safe, secure, and satisfying relationships with others based on intimacy, equality, and commitment. When they are lacking the child will have a difficult time getting along with others at home and at school and will be at risk for a diagnosis of ADHD.

- ***Empathy:*** A child learns to express empathy for others when he has received consistent caring and consideration from his own caregivers. If he has not experienced caring, he might be capable of hurting others without apparent remorse. Exacting revenge on others might become more important than making friendships.
- ***Trust:*** A child learns to trust others when primary caregivers are accessible, responsive, and benevolent. When caregivers are unavailable, unpredictable, or abusive, the child does not turn to them for soothing or reassurance when he or she is afraid or in pain. Unable to rely on adults to meet his needs and keep him safe, the child becomes his own inadequate caregiver and protector.

• ***Affection:*** A child learns to express affection openly and to accept it from others when he experiences comfortable closeness with primary caregivers. A child diagnosed with ADHD is usually reluctant to offer or accept nurturing physical contact with his parents or his peers. He may resort to pushing or hitting as a way of attracting attention to himself or conveying interest in others.

• ***Reciprocity:*** A child who experiences satisfaction and gratification in the daily give-and-take with primary caregivers will learn to be considerate, playful, and fair in his interactions with others. Children who fail to learn reciprocity from their caregivers may be selfish, inconsiderate, or coercive in their social interactions.

• ***Expression:*** A child learns to express his or her real self when he or she receives encouragement from caregivers and when he or she develops in an environment where it is safe to be vulnerable. Children diagnosed with ADHD are usually afraid of their own feelings and reluctant to acknowledge or express them to others. They may be willing to accept considerable pain rather than to risk emotional vulnerability.

• ***Respect:*** A child learns to respect others when he or she is valued and encouraged by primary caregivers. Children who fail to receive respect from primary caregivers do not learn that they are worthy of respect and are likely to make a habit of disrespecting others. Children diagnosed with ADHD rarely show respect for their parents or adult authority figures.

In our model for explaining the origin of ADHD, these two main categories of attachment deficits (self-regulation and relating skills) are seen as causing the behaviors that drive parents, teachers, and therapists to distraction. We consider them to be treatable problems that develop after conception, the results of traumatic bonding breaks between caregiver and child—not as the permanent symptoms of a genetically transmitted disease.

The Symptoms of ADHD

The three main symptoms of ADHD in children are thought to be hyperactivity, inattention, and impulsivity. In addition to these primary indicators, there are a number of annoying habits and behaviors which are considered to be diagnostic criteria for this problem. For example, parents and professionals agree that children with ADHD are often defiant, forgetful, irresponsible, oppositional, intrusive, disruptive, manipulative, careless, disorganized, oblivious, and obnoxious. They usually do not get along well with other children. When it comes to identifying the causes for these symptoms, however, there is little agreement as to which of these diagnostic criteria are attributable to genetic defects and which to developmental deficits.

Hyperactivity

This is the symptom that distinguishes ADHD from simple ADD (attention deficit disorder), but there may be as many possible explanations for hyperactivity as there are children. In the newborn, for example, hyperactivity could be the result of prenatal or perinatal trauma. In the preschooler, it could reflect neurological anomalies in his prefrontal cortex—the result of insufficient soothing interactions with his caregivers. In a 10-year-old, it could be a symptom of chronic autonomic hyperarousal. Children who have been traumatized by abuse, neglect, or insecure attachment develop brains that are hypervigilant to potential threats in their environment. The longer they remain stuck in the "fight-or-fight-or-freeze" response, the more likely they are to experience adaptive changes in their central nervous systems and maintain a baseline state of physiological arousal and hyperactivity (Perry, Pollard, Blakely, Baker, & Vigilante, 1995; van der Kolk, 1994).

Impulsivity

Working with dozens of children diagnosed as having ADHD, we have learned that most of them exercise very little control over their impulses. They seem to share the same immature belief: "When I really, really want something, I should not have to wait for it!" Lack of empathy for others could be a partial explanation for a child's pattern of impulsive, intrusive behaviors. This symptom could also be the result of neurological deficits in that region of the frontal cortex that is responsible for thoughtful control of behavior. Furthermore, when a child is experiencing a persistent state of anxiety and autonomic arousal, he or she will have a difficult time accessing the part of his thinking brain that would allow him or her to stop and consider the possible consequences of his choices and behaviors (Perry et al., 1995).

Inattention

This has always been the diagnostic cornerstone for ADD and ADHD. It will continue to be a controversial notion until we have conclusive evidence proving that some children are inherently incapable of paying attention as well as others. Our own experience with children diagnosed as having ADHD tells us they can pay attention as well as other children when they are properly motivated. We have also become convinced that children diagnosed with ADHD are only motivated by things that are stimulating or gratifying in some way. It is as though they have become addicted to stimulation. They seem to all share a common belief: "Anything that is not stimulating or pleasurable should be ignored." This pattern makes sense when you consider that they have grown up in an environment in which they were overstimulated and undersoothed because of bonding breaks. They have become like little cars without brakes, with their accelerators stuck to the floorboard.

It is interesting to note that Russell Barkley, a long-time researcher in the area of ADHD has, in his most recent book, called for a new model to explain the etiology of this disorder. He makes the point that, despite its being consistently reported in parent and teacher ratings of problem children, a measurable deficit in attention has not been found in children diagnosed as having ADHD (Barkley, 1998).

The Cycle of Conflict between Caregiver and Child

Children with attachment deficits spend much of their time locked in a constant power struggle with a primary caregiver because the thing they desire most (emotional closeness) is also the thing they fear most. Consider, for example, the cycle of conflict that usually exists between a child diagnosed with ADHD and his caregiver and how this pattern is the result of his trying to make a connection without having the skills necessary for healthy attachment. As can be seen in Fig. 2, the cycle begins when the child is experiencing a strong negative emotion such as anger, sadness, loneliness, or fear. Lacking the capacity for self-soothing, impulse control, and expression, the child attempts to connect with the parent through intrusive, demanding, attention-seeking behaviors. The parent begins to feel irritation and resentment but is unable to express empathy, affection, or respect for the child and responds by criticizing, threatening, or hitting him or her. The child reacts by tuning the parent out and silently planning his or her revenge or he or she becomes defiant and coercive and raises

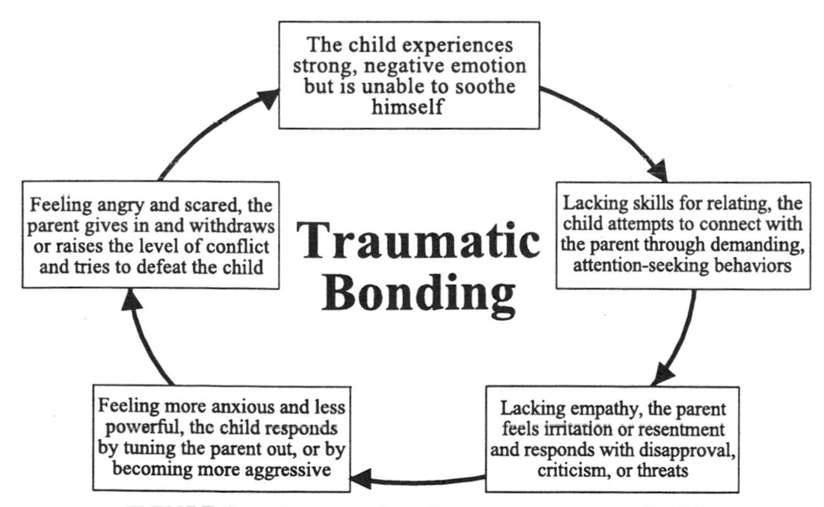

FIGURE 2. The cycle of conflict between parent and child.

the level of his acting-out behaviors. At this point the parent, feeling angry and scared, either gives up and withdraws or raises the level of conflict in an effort to defeat the child. After each fight, both the parent and the child are left frustrated and angry and determined to get even by winning the next round. Together, they have created a habitual pattern of traumatic bonding, a relationship based on conflict rather than intimacy.

Eight-year-old Ashley understood the cycle of conflict. One day, during family therapy, she explained it to her mother: "Mom, you need to let me alone when I'm upset. I can handle that problem myself. You should go calm *yourself* down. Hitting me doesn't help!" Despite her apparent understanding of this problem, Ashley acted out the same pattern of escalating conflict at school where she was eventually punished for hitting her best friend.

A MODEL FOR TREATING THE FAMILY OF A CHILD WITH ADHD

Whenever possible, the treatment-of-choice for a child diagnosed as having ADHD should be family therapy. This is especially true when the child is between the ages of 5 and 15. The reason is simple: children with ADHD are not likely to make significant changes in their thoughts or behaviors without simultaneous changes in their family systems. Even in the case of a hyperactive and impulsive child entering a new and relatively healthy family system, such as a foster or adoptive family, that system will require extremely competent and effective parenting skills in order to succeed with such a child. We have developed a six-step model for family therapy based on corrective attachment theory, family systems theory, cognitive and behavioral techniques, and some exciting things we have learned recently about the "rewiring" of a child's brain.

Engaging the Primary Caregivers as Clients and Cotherapists

The most critical prerequisite for appropriate, successful treatment of the child diagnosed with ADHD is the cooperation of the adult caregivers. It is absolutely essential when the primary caregivers are the biological parents because they have probably contributed, in some ways, to the development of the child's behavioral and emotional symptoms. It is also necessary in foster and adoptive families. All children who exhibit attachment deficits need to experience a positive connection with a willing and motivated adult caregiver in order to become whole. The therapist's role is that of a "relationship coach"—not a "surrogate parent."

Overcoming Parent Resistance

Unfortunately, a therapist will frequently encounter major resistance when asking for the trust and cooperation of primary caregivers. The key to overcoming parents' resistance is to listen carefully to what they are saying and feeling and to look for ways to empathize and join with them. Until caregivers feel heard and understood, the therapist should not begin to focus on their contributions to the child's problems.

The Exhausted, Overwhelmed Parent

By the time they arrive at a therapist's office, many caregivers for children with ADHD are mentally and physically fatigued and overwhelmed by negative emotions. They are feeling blamed and criticized by family, neighbors, teachers, and other professionals who have said things like: "There's nothing wrong with that kid that the right kind of parenting wouldn't fix!" These parents need to know that someone believes they have been doing their best for their child. They need consistent empathy and compassion from a therapist in order to cope with their own feelings of frustration, hyperarousal, and hopelessness.

The Medication-Seeking Parent

These caregivers have become convinced that ADHD is a genetically transmitted medical problem and that their child will have to take psychostimulant medication in order to be "normal." They may be unwilling to cooperate with family therapy until medication has failed to produce the desired result for their child. One of our clients brought in her 10-year-old son for an assessment of his impulsive and aggressive behaviors, saying: "I want a professional opinion to determine whether or not he has ADHD and needs medication." The boy's psychosocial assessment revealed that his biological father had abandoned the family when the child was 4 weeks old and had never seen the boy again. Furthermore, it became clear that the boy was being criticized, humiliated, and abused in a habitual way by his stepfather. When the mother received the recommendation that the entire family should begin therapy, she adamantly insisted on a referral for her son to see a psychiatrist. "But I want to start at the top—with the doctor! If he rules out ADHD, then we'll consider other treatments. Why waste time with therapy if he just needs medication?" The child was taken to a psychiatrist who prescribed Ritalin "to help him control his anger." The mother apparently believed that the physician had ruled out the necessity for family therapy. Three months later, she telephoned to say that his aggressive behaviors had escalated to the point that he had been suspended from school for hitting his classmates. From this incident one might conclude that, whenever possible, a therapist should refer this parent to a physician who will emphasize the importance of family therapy to accompany the use of medication.

The Guilty Parent

Caregivers may feel incredible guilt for some surprising reasons, such as for secretly disliking their child diagnosed with ADHD or for passing on to him their so-called "bad genes." Caregivers who harbor anger and resentment for their child may need help in finding ways to express their negative emotions, without blaming or shaming that child. Because guilt is a counterproductive emotion and an obstacle in the process of engaging the parents, a therapist should assess for feelings of guilt and help the parents to assuage these negative emotions through education and cognitive restructuring. The therapist should reassure them that they are the people best-suited to help solve their child's problems.

The Hopeless and Helpless Parent

Some caregivers will have already tried a variety of parenting techniques and behavioral interventions without any success and will be very discouraged. They may be convinced that: "We have tried everything possible and nothing works with this kid!" They can be very pessimistic about the usefulness of family therapy. It usually turns out that these caregivers have been inconsistent and irresolute with their discipline and have given up without much of a struggle. A therapist should help the parents understand why their efforts have been unsuccessful and teach them to be creative and persistent enough to gain control of their family. Have them start with limited, sequential goals that can be easily attained.

Parents Who Need Their Child to Be "Ill"

Some caregivers need to believe that their child with ADHD has a medical disease and is "ill." They may cling to the diagnostic label as though it somehow absolves them from having made the slightest contribution to their child's problems. They may be reluctant to participate in family therapy without a formal acknowledgment of their child's medical pathology. Because their greatest fear is that they will be blamed and identified as "bad parents," these caregivers require plenty of reassurance from a therapist. They may need help understanding their own needs and goals as parents and maintaining hopeful expectations for their child in treatment.

The Narcissistic Parent

A sizeable percentage of the biological parents of children diagnosed as having ADHD appear to be so self-focused that they experience very little empathy for their children. One such mother of an encopretic 12-year-old boy, diagnosed with ADHD, had become so frustrated and angry with him that she literally could not stand the sight of him. She expressed to us the extreme dread she felt when he arrived home from school each day. When asked to guess what he was feeling after being taunted and ridiculed by his peers at school, she responded with hostility: "Oh he doesn't even notice when they call him names. Nothing bothers this kid!" Truly narcissistic parents are extremely difficult to engage in family therapy because they are so angry at their children. Furthermore, they usually do not tolerate suggestions or corrections from a therapist. The task of engaging such parents as cotherapists will require the expression of genuine concern for their suffering and their plight as well as the consistent modeling of empathy for their child. When working with narcissistic parents, the therapist should be aware that his or her own countertransference issues might interfere with his ability to help them.

The Grieving Parent

Many caregivers for children diagnosed as having ADHD are actively grieving the loss of their expectation for a normal and healthy child—one that would reflect positively on them and their parenting. They have lost the "perfect child" and feel powerless to cope with the "real child" in their home. Our experience has taught us that parents' feelings must be brought to the conscious level and expressed before we can expect them to relate differently with their child. This has proven true whether the child is their biological offspring or adopted. In either case, the parent has harbored a secret longing for a particular relationship with this child—a relationship that has not developed. Until the parents identify and accept their true feelings about the child, they will probably act out their anger against the child who has disappointed them. Engaging these parents requires that a therapist help them through the stages of the normal grieving process before they are able to accept the reality of their situation and relinquish their lost fantasy.

Polarized Parents

Some pairs of caregivers have become thoroughly polarized and are so focused on blaming one another that they are unable to work on solving the child's problems. This is especially true in families where parents have experienced a bitter divorce. In some cases, this polarization can be viewed as a reaction to the child's disturbed behaviors. In others, preexisting conflict between the parents may be a major cause of the child's problems. The therapist will need to contract with such parents for some basic elements of healthy coparenting in order to be successful with their child. Some couples will need marital therapy before they can truly cooperate with treatment for their child.

Ambivalent Parents

Some foster and adoptive parents have already given up on their child diagnosed as having ADHD and come to therapy secretly hoping that treatment will be unsuccessful so they can justify returning the child to the agency from which he or she came. Refusing to cooperate with family therapy might represent their unconscious attempt to sabotage their child's placement. Occasionally, birth parents also will be experiencing such extreme frustration that they are prepared to give up custody of their child to the state. This constitutes a very grim situation—one in which the therapist must stop and help the parents make the decision that will be in the long-term best interest of the child.

"Triggered" Parents

Some caregivers for a child diagnosed as having ADHD live each day of their lives in emotional upheaval because their child's behavior is reminding them of painful events from their own childhood. One mother came in with a 9-year-old boy and his 6-year-old sister, both diagnosed as having ADHD. During the initial assessment, while describing her son's aggressive treatment of her daughter, she burst into loud, tearful wailing: "I never wanted a boy! I was so afraid he would turn into a little bully like all those boys who tormented me in school, and now look at him! That's exactly what he is—a bully! I just hate him!" This was a parent who could not be engaged in family therapy until she had worked personally for a while in group therapy. Luckily we had the option of involving the father more actively in the parenting process while the mother concentrated on disarming the "triggers" from her own childhood. The task of becoming a good cotherapist for one's child requires that the parent offer that child a new and improved relationship.

Engaging the Parents as Cotherapists

The first task for a therapist who is beginning family therapy with a child diagnosed as ADHD is to meet with the parents and tell them flatly that they will be the most important players in the process—that no therapist has enough skill to succeed with their child without their wholehearted and dedicated assistance.

Meet with the Parents by Themselves

It is important to exclude the child from the first meeting with his caregivers. They need to know that the therapist is aligning with them and the child also needs to understand that fact. It may be helpful to start by giving them permission to verbalize their feelings of anger, frustration, and disappointment while explaining that they will have to temper those feelings with considerable empathy when they express them in sessions with their child. Explain the notion that every family can be accurately viewed as a unique and distinct system in which the actions and reactions of one member influence the actions and reactions of every other member. Give the parents a message of hope—not of resignation: "The bad news is that you might have contributed to his problems in some way. The good news is that there is a lot you can do to help him."

Collect Biopsychosocial Information about the Parents

Look closely at the parents' developmental histories, especially the emotional traumas from their own childhoods as well as the relationships with their own primary caregivers. Help them to discover the specific attachment patterns they learned from those individuals. Identify and explain all issues that might be interfering with their ability to successfully parent this particular child. Use this information to raise their level of self-awareness and to increase their empathy, tolerance, and consideration for themselves and for their child.

Obtain Details of the Child's Prenatal and Postnatal Development

Learn everything you can about the first 2 years of his or her life. Explain the cycle of human attachment as it normally occurs for a healthy child in the first year of life as well as the predictable consequences of a child's failure to learn trust and self-soothing. Identify incidents of childhood trauma and frame them as bonding breaks that have caused important delays in their child's development. When parents come to understand the significance of unavoidable breaks in the bonding with their child, it may give them permission to feel less guilty. They may experience feelings of relief with the realization that: "There are logical explanations for the problems my child is having. Perhaps I did not give birth to a demon after all!" The therapist should help parents to understand that, although their skills may be adequate for raising the average child, they will need to become "expert" parents in order to succeed with the child they have now.

Outline the Basic Goals and Objectives of Treatment

Explain the priorities and the sequence of the treatment plan. It will be important to spell out specific expectations for the parents, both at home and in the therapy sessions. For example, in the area of parent–child communication, we suggest that the therapist should contract with the parents to avoid all shaming interactions such as hitting, yelling, name-calling, criticism, and sarcasm. They should have concrete, specific alternatives for the habitual negative interactions they have been experiencing with their child. Give the parents a written copy of the individualized treatment plan, emphasizing the main areas of focus for their particular child.

Listen Carefully to the Parents' Concerns

Allow them as much input as possible in the creation of the treatment plan and ask for their input and feedback at every turn. They might need help maintaining their objectivity but, after all, they know more about their child than anyone else does. Pay close attention to the efforts they have already made in trying to solve problems and encourage them to figure out why these attempts have failed and what they might do differently.

Obtain their input in the setting of realistic treatment goals for their child and work for consensus when possible. If the therapist fails to listen to the parents in good faith, they are not likely to accept the job of cotherapist in good faith.

Teach the First Rule of Good Parenting: "Take Care of Yourself"

Encourage the parents to make sure they find time for their own recreation and relaxation. Explain to them that one of the keys to mental health is maintaining a balanced lifestyle. Urge them to develop more empathy for one another so that they can avoid their blaming mode and more easily access their problem-solving mode. Remind them that a child learns to take care of himself physically and emotionally by relating to adults who take proper care of their own mental and physical health.

Formulating a Detailed Assessment of the Child's Problems

The therapist's next task is to collect current information about the child diagnosed as having ADHD from parents, grandparents, siblings, and teachers and somehow reconcile that data with his or her own observations and interactions with the child. It is important to maintain a healthy skepticism regarding anecdotal reports from anyone who is really angry with the child, especially parents and teachers. Complete a thorough assessment of the child by obtaining the answers to three simple questions.

What Is This Child Doing?

The therapist must develop a thorough and accurate inventory of the child's aversive behaviors and a reliable estimate of his ability to control them. A careful assessment will also include the determination of whether the behavioral problems occur mainly at home, school, or equally at both. If the problems occur mainly at home, it is important to know whether the child acts-out more in the company of his mother, his father, or equally with each parent. If the problems are mainly at school, it is important to determine whether the child is acting-out because he does not relate well with peers or teachers or because he has he is unable to succeed academically. The child's aversive behavior will consistently reflect his system of primitive defenses. He will usually rely on habitual mechanisms such as denial, minimization, avoidance, and blaming to protect him from painful feelings. At this point in the therapy process, the caregivers can practice reacting differently to the child's aversive behaviors, armed with the understanding that his behaviors are driven by his negative feelings, which are the products of his or her distorted thinking.

What Is This Child Feeling?

The next step in the process of developing a thorough assessment of a child diagnosed as having ADHD is the identification of the painful feelings being acted-out as aversive behaviors. The list of negative emotions will always include significant levels of mad, sad, and scared along with some others like loneliness and worthlessness. The child, however, will usually not admit to having these feelings because he fears that will make him vulnerable and deprive him of the power and control he has worked so hard to get.

It is also important to establish a valid estimate of his capacity for self-soothing, self-expression, and other forms of affect-modulation. The child's history provides information about his or her developmental trauma and should reveal the causes of his or her anger, the reasons for his or her sadness, and the sources of his or her fear. The process of identifying the specific areas of mad, sad, and scared will provide a trail of breadcrumbs to the cognitive distortions responsible for the child's painful feelings.

What Is This Child Thinking?

A child who has failed to form a secure attachment with a primary caregiver does not think the same thoughts as a healthy child. Distorted thoughts and delusional beliefs are predictable sequelae of bonding breaks and attachment deficits in an insecurely attached child. For example, when a child fails to learn *trust* in the first 2 years of life he may develop core beliefs such as these:

"Adults are unreliable, unresponsive, and untrustworthy."
"I must control others in order to be safe."
"Being close to others is not pleasant."

A child who does not receive the modulating responses he or she needs from a primary caregiver fails to learn self-regulation and may hold these beliefs:

"I am not able to control myself."
"When I really want something, I should not have to wait for it."
"Things that are difficult should be avoided."

A child who lacks sufficient soothing interactions with caregivers does not learn self-soothing and may become convinced that:

"Feelings are dangerous and should be avoided."
"Things that are not stimulating or pleasurable are a waste of time."
"Being alone really scares me, but I cannot afford to let anyone know that."

When a therapist has identified the cognitive distortions that underlie the painful feelings and drive the aversive behaviors of the child with ADHD, he or she is ready to work out the specifics of an individualized treatment plan. The child can develop new beliefs from his new experiences with the caregivers—cognitive restructuring that fosters healthy attachment between parent and child instead of the traumatic bonding that has occurred in their cycle of conflict.

Helping Caregivers Provide a Secure Family Environment and Remedial Parenting

In order for family therapy to succeed with a child diagnosed with ADHD, the caregivers must work together to create a healthy environment—one that will provide the safety, security, and structure required for repairing attachment deficits. Likewise, this ambitious and complicated task will require a set of consistent parenting strategies that will have the effect of containing the child in a sturdy box with a cushioned lining.

Establishing a Secure Family Environment

Many children who have experienced bonding breaks and attachment deficits, and have developed symptoms of ADHD, grew up in chaotic homes where life was unpredictable and the environment unsafe. This explains why these children work so hard to control everyone and everything in their lives. In order for family therapy to succeed with a traumatized child, his or her family environment will have to be safe enough for him or her to give up his need for control.

Eliminating All Hitting, Yelling, Criticism, and Sarcasm from Family Interactions

A child who has been traumatized by bonding breaks usually possesses a painfully negative self-image that is being maintained by cognitive distortions and feelings of self-loathing. When a caregiver hits, injures, or abuses a child who dislikes himself, that child's belief that he or she is bad is strengthened, and he or she becomes more likely to do bad things. Because he or she hates being reminded of his or her worthlessness, this child becomes angry and vindictive or tunes the caregiver out when he or she is subjected to verbal abuse. He or she will respond to the slightest criticism with aggressive behaviors (hyperarousal) or may freeze and tune out his or her antagonist (dissociation). The caregiver who tries to control such a child with criticism or intimidation has no chance of helping the child and can only exacerbate the problem. Parents must learn to be *calm and firm* when correcting their child, but *loud and excited* when praising him or her. In order to help this child, every attempt to interact with him or her must be framed in empathy. Parents may need to be reassured that the therapist will help them develop more effective, nonthreatening ways to manage the child's aversive behaviors.

Creating a Warm Family Climate Characterized by Empathy, Affection, and Respect

One of the primary symptoms of the child with ADHD is the lack of relational skills. Because the child has been traumatized by bonding breaks, his neurological development has been vandalized and his brain is not programmed for expressing or modulating feelings of love and affection. He is, however, capable of learning the building blocks of attachment when they are modeled by primary caregivers. When a child has the opportunity to feel understood, cared about, and respected on a consistent basis, he or she usually learns to feel those emotions for others and for him- or herself. Thousands of safe, predictable, and empathetic interactions with caregivers have the cumulative effect of helping the child build a realistic, positive self-image.

Establishing Clear and Consistent Rules, Roles, and Routines for Every Member of the Family

Every child needs fair rules and expectations, clearly promulgated and consistently enforced, in order to feel safe and secure in his family environment. He or she needs to have a thorough understanding of his or her role and responsibilities in the family and to trust that the expectations imposed on him or her by his or her caregivers will be realistic and benevolent. Consistent routines and schedules for meals, chores, television, getting up in the morning, and going to bed at night help him or her to learn responsibility, self-discipline, and obedience.

Maintaining Effective Patterns of Communication between Family Members

The child diagnosed with ADHD usually has a difficult time expressing him- or herself. Lacking verbal skills, he or she tends to rely heavily on nonverbal cues such as facial expressions and body movements to know when he or she is in danger. However, our experiences with children diagnosed as having ADHD teach us that they are likely to interpret neutral responses from peers and adults as being negative and threatening. Healthy family communications are honest, clear, calm, and respectful. Adult caregivers usually set the tone and tenor of communication within a family, not only in terms of what they say and what they do not say, but also in terms of how their messages are delivered. Caregivers should be aware that the absence of verbal communication sends a nonverbal message to their children. Children with attachment deficits generally interpret nonverbal communications as being negative. When caregivers communicate with their children in ways that are warm, loving, and accepting, they increase the opportunity for emotional attunement, trust, reciprocity, and attachment.

Modeling Self-Care as a Personal Responsibility

In families where a child is diagnosed with ADHD, conflict often erupts when that child is asked to be responsible for his or her own self-care in dealing with such things as homework, punctuality, and personal hygiene. Caregivers who take care of themselves set a good example and model healthy responsibility for their children. A child needs to learn that each of us is responsible for guarding his own mental and physical health. "Putting oneself first" is a practical idea, not a selfish one. It is another way of saying: "When I am meeting my own needs for healthy recreation, relaxation, and gratification, I will be better able to protect you and see that you have everything you need to be strong, happy, and successful."

Principles of Remedial Parenting

Children who display the behavioral and emotional symptoms of ADHD require first-rate parenting in order to overcome the effects of bonding breaks and attachment deficits. They need to experience consistent application of remedial strategies and techniques to constrain their aversive behaviors and emotions and help them to internalize healthy ones. Once again, the therapist's role is that of a teacher for the parent–cotherapist.

Teaching the Child Discipline without Punishment

Punishments are arbitrary, mean-spirited, and toxic to the bond between parent and child because they reinforce traumatic bonding, lead to negative power struggles, and create a wish for revenge. Effective parenting teaches the child diagnosed with ADHD which behaviors are acceptable in given situations and explains what consequences he or she will experience when he or she does not comply with those expectations. Consequences differ from punishments in that they are administered in a calm, considerate, businesslike manner. They are always logical and benevolent. Giving the child an opportunity to make choices regarding his behavior can help him learn self-examination, self-confrontation, and self-determination. Parents must be careful, however, to offer the child only those choices that the parents can easily live with. When unacceptable behavior is presented to the child as an available option subject to consequences, that child might eventually decide that as long as the the penalties can be tolerated, he or she is free to do whatever he or she wants.

Requiring Obedience Responsibility, and Respect

The child with attachment deficits always wants to control others and is reluctant to accept a parent as his "boss." Attempts at gaining control of the family usually involve deliberate, provocative behaviors that are disobedient, disrespectful, and irresponsible. One of the most important things a child with ADHD must learn from his parents is that obedience is not optional. Every child needs a good boss to teach him how to behave in all of life's situations. Parents who consistently demand obedience in ways that are empathetic, tolerant, and considerate, and who back up their demands with logical consequences, will eventually persuade the child to relinquish his need for control in exchange for the protection of a strong, benevolent boss.

Establishing Structure, Limits, and Boundaries

A child whose early experiences have been characterized by chaos and unpredictability usually lacks impulse control and experiences life in a chronic state of hyperarousal. Impulsivity causes daily interactions with others to also be unpredictable, thereby increasing his or her frustration and anxiety. A child who lacks impulse control normally benefits from the application of clear and consistent boundaries in his or her daily routine. Anchored by the predictability of limits and structure, the child is able to feel secure enough to begin internalizing self control. A child learns to modulate emotion and behavior when the parents monitor and reinforce his or her compliance with prescribed boundaries, both physical and relational.

Parenting That Is Proactive, Not Reactive

A reactive style of parenting is especially counterproductive with the child who has ADHD because it reinforces his or her belief that the world is unpredictable and unsafe. It increases his determination to gain control of others.

Reactive parenting exacerbates the frequency and intensity of power struggles between parent and child. Proactive parenting involves the thoughtful anticipation of circumstances in which the child might experience emotional hyperarousal and provides the child with choices for appropriate decision-making that can help to modulate his or her feelings and lead to self-soothing. The proactive parent avoids power struggles by choosing the time, place, tone, and pace for all interactions with the child.

Contracting for the Child's Cooperation

Children with attachment deficits and a diagnosis of ADHD often lack the capacity for reciprocity and synchronicity in their interactions with others. Parents can work to remediate these deficits by contracting with their child for the attainment of specific behavioral and relational goals that are based on expectations appropriate to the child's developmental level. When mutual satisfaction results from the successful performance of a contract, it can build trust and respect between a parent and a child.

Promoting Interaction, Attunement, and Attachment between Caregiver and Child

This is perhaps the most important element in our model for family therapy with a child diagnosed as having ADHD. This is the part that requires the therapist to be most active and to employ the highest skill. It is here that the therapist uses his knowledge of human attachment and attaching behaviors to help caregiver and child form a healthier bond that will be the foundation for their new and improved family.

Holding by Therapist and Parent

If the child's history includes a pattern of inappropriate or hurtful physical interactions with adults, both the parent and the child will benefit from learning specific ways to use physical contact to promote affection and trust. In severe cases, when a child has been diagnosed with a serious attachment disorder, holding therapy with a specially trained therapist might be required. Children who have been identified with lesser problems (attachment deficits) can benefit greatly from being held by their parents. The parent, with the joint roles of caregiver and cotherapist, is the best person to teach the child that healthy touch can soothe his or her anxiety and alleviate his or her fear of being alone. The therapist teaches the parents how to hold the child in appropriate ways. Some children may need cradling, stroking, and rocking although their chronological age might suggest they are beyond that. Being held on a parent's lap, breathing in synchrony with the parent, listening to lullabies, or just gazing into the parent's eyes—all of these things offer the child tangible experience of soothing and allow him to imagine how it might feel to soothe himself. Surprisingly, few children will refuse an offer to be held by their parent in the therapist's office. Therapeutic holding allows the child to experience safety and security within the context of closeness and physical contact. It also helps to build reciprocity and repair attachment deficits.

Videotaping Parents and Child in Therapy Sessions

It can be very helpful to videotape sessions in which a caregiver and child interact in a significant way. Schedule time to review these videotapes with parents, using the rewind and slow-motion buttons on the VCR. Therapist and parents work together to identify important elements that might have been overlooked when viewing the action in real time. For example, parents will often be astonished to see how many of the child's cues and expressions they had failed to observe or acknowledge. This process can help parents to improve their communications with the child, both verbal and nonverbal, and enable them to see, understand, and correct their own faulty relational behaviors.

Teaching the Parents to Express Empathy for the Child with Every Interaction

A child who exhibits behavioral patterns of hyperactivity and impulsivity usually receives a constant stream of criticism, contempt, and negative commands from parents, teachers, and others. Eventually, the child sees himself not only as doing bad, but as actually being bad. When he acts out the bad feelings as bad behaviors he completes the self-defeating cycle by generating more criticism and contempt from his caregivers.

The most powerful tool available to the parents for breaking this cycle of acting-out behaviors is their expression of empathy for the child. In order to provide empathy for him or her on a consistent basis, the parents must learn to separate their feelings about the aversive behaviors from their feelings about the intrinsic value of their child, responding to him or her with understanding and acceptance even when he or she is trying to provoke them. The therapist can help by interpreting the bad behaviors as the child's misguided efforts to connect with his or her parents without risking rejection. When the child realizes that his or her parents understand and accept his or her feelings as appropriate responses, that child can begin to give up the behaviors he or she used to avoid those feelings. Their empathy helps the child accept what he or she has been feeling and gives permission to change what he or she has been doing.

Encouraging the Child to be Affectionate on the Parents' Terms

Children diagnosed with ADHD, although they might express physical affection to peers and family members, are usually reluctant to accept nurturing, physical contact initiated by others. Sometimes they are labeled as "sensory defensive" because they exhibit a startle response when touched without warning. The therapist looks for opportunities to encourage the child to tolerate situations in which he cannot control the expression of affection. The parents will need to be patient, but persistent, taking the lead in offering the child affection that is nonthreatening. As the child learns to accept closeness and nurturing physical contacts his or her baseline of anxiety will be lowered and he or she will be able to relinquish his or her need for control.

Demonstrating and Teaching Playful Interactions between Parent and Child

Early bonding breaks rob caregiver and child of the opportunity to fall in love with one another and inhibit the development of secure attachment. The therapist can address these relational deficits by teaching the caregivers how to interact with their child in the play therapy room. These structured, choreographed interactions help the caregivers and child to feel good about having fun together and to learn important qualities like patience, perseverance, initiative, and sportsmanship. The caregivers learn to be more sensitive and more responsive to the child's needs and to be more emotionally available to the child without frustrating him or her or intruding in his or her space. These are important steps in the development of attunement, synchronicity, and reciprocity.

Accessing Collateral Therapies to Increase Self-Regulation and Relating Skills

We have identified a variety of adjunct therapies, some new and some well known, which can be included in a treatment plan to help a child learn self-regulation and relating skills and to promote attaching behaviors in parent and child. Parents, as cotherapists, should be involved in making decisions about creative options for treating their child. Each of these modalities is considered to be appropriate for treating one or more of the symptoms of ADHD. They are all compatible with our model for explaining the origin of ADHD because they serve to heal attachment deficits by rewiring the neurology of a child's brain in one way or another.

Group Behavior Therapy

Group therapy has always been a wonderful place for children to learn self-regulation and relating skills. The artful therapist uses games, rituals, and role-plays to teach children how to control impulses, give verbal expression to feelings, accept delayed gratification, and practice interacting with peers. Likewise, the child will benefit from the structure of the group rules and from the experience of exercising control of his impulses for the good of the group (Halperin & Kymissis, 1995).

EEG Biofeedback

Also known as neurofeedback, this treatment option has been used in one form or another for more than 20 years. A current version of the procedure uses a computer video game in conjunction with encephalographic equipment to teach a child to recognize when his or her brain is in a particular state and to actively enhance or suppress that state by focusing his or her thoughts, thereby improving his or her chances of success with the game. Treatment goals include the enhancement of cognitive skills, such as attention, concentration, and academic performance. This protocol is thought to require a minimum of 20 30-min sessions for successful outcome. One major drawback with EEG biofeedback is its cost, which may be prohibitive for many families (LaVaque & Rossiter, 1995).

Eye Movement Desensitization and Reprocessing (EMDR)

This is a revolutionary psychotherapy procedure that has become widely accepted during the past 5 years, especially in the treatment of traumatized adults and children. It is being used primarily to treat symptoms of anxiety, anger, guilt, and Post-traumatic Stress Disorder. The basic procedure requires a specially trained and certified therapist to guide the client in concentrating on a traumatic memory or emotion while moving his eyes back and forth in rapid movements. Although we do not have consensus for a precise explanation of the healing dynamic in EMDR, it is assumed that when the rapid eye movement coincides with the client's emotional reexperiencing of the trauma, the brain's processing and interpreting of specific memories is altered and hyperarousal is diminished (Forrest & Shapiro, 1998).

Thought-Field Therapy (TFT)

This is another revolutionary therapy that has become quite popular during the past several years. Thought-field therapy is being used by specially trained and certified therapists to treat anxiety, fear, rage, grief, Post-traumatic Stress Disorder, and other symptoms of hyperarousal. The treatment is designed to clear blockages in a person's energy flow that are the result of disturbing thought patterns derived from traumatic experience. The procedure requires the client, at the direction of the therapist, to tap on his upper body at specific places (accupressure points) with his or her fingers or hand in a specific sequence (algorithm) while thinking about a targeted problem. The treatment is believed to approximate the techniques of acupuncture, without the needles, because it focuses on the classic meridians and energy points used in that ancient healing discipline (Callahan & Callahan, 1996).

Brain Gym

Also known as Educational Kinesiology, this relatively new treatment program has been promoted primarily as a benefit to cognitive processing in children with learning disorders. It is also being used to treat other specific deficits in children and adults, such as hyperactivity, behavioral dyscontrol, and relationship problems. Brain Gym is a highly structured program of physical movements and activities, sequenced to meet specific needs, and designed to integrate the left- and right-brain functions, thereby improving performance in complex human tasks such as organizing, remembering, concentrating, and communicating. Children who are traumatized by bonding breaks appear to get stuck in a hyperaroused feeling state controlled by the right side of their brains. Consequently, they have a very difficult time accessing the left-brain functions required for academic success, such as remembering, organizing, and expressing themselves verbally. Not surprisingly, Brain Gym practitioners report success in reducing the symptoms of hyperactivity and impulsivity in children diagnosed with ADHD (Dennison & Dennison, 1994).

Art and Play Therapy

In addition to their value in the process of assessing children, these therapies may also help the child who has trouble expressing his emotions verbally and has been acting them out as aversive behaviors. Some children appear to derive great benefit from drawing or from role-playing their feelings in the presence of an adult who can reflect them back in the form of words. This mirroring of affect is likely to be something they have lacked during the years they needed it for emotional and neural development. Art therapy or play therapy, when parents are included as cotherapists, may help to break the cycle of negative interactions with their child. When a child learns that words can be used to gain understanding, empathy, and acceptance from caregivers, he will begin to respond differently in situations where the aversive behaviors had been creating problems for the family (Norton & Norton, 1997).

Martial Arts, Chess, Dance, and Organized Sports

This is a short list of structured activities that can help a child diagnosed with ADHD to learn self-discipline and impulse control. They all require a child to exercise control of his or her mind and body together within the context of specific rules and movements. They are also challenging and stimulating and require consistent concentration and perseverence for mastery. These are opportunities for a child to experience internal and external rewards for hard work. They also require him or her to interact with peers in ways that promote cooperation and reciprocity. The more effort the child applies to the task at hand, the more attention and approval he or she receives. The better he or she learns the rules and tries to follow them, the better his or her chance for success and feeling good about him- or herself.

Suggesting Psychotropic Medications as a Last Resort

We believe that children are programmed for health and have a tremendous capacity for self-healing. For this reason, interventions that promote permanent internal healing for children, like family therapy, are preferred over temporary drug-induced relief that does not bring about real change. More importantly, medications that maintain a child in a state of lethargy or euphoria can be major obstructions in the course of family therapy. There are times, however, when a therapist's best option might be to encourage caregivers to seek medication for their child or to continue its present use.

If, for example, a child's behavior is so outrageous that he or she cannot be managed at home or at school, psychopharmacologic intervention might be advisable as a temporary solution to a crisis. When caregivers are unwilling or unable to participate as cotherapists in family therapy, medication might be recommended for the longer term. Furthermore, most children diagnosed with ADHD who are brought to a therapist's office are already taking psychostimulants, antidepressants, or antihypertensives to control symptoms of the disorder. A therapist should not try to pressure his cotherapists to abandon a treatment plan established by the child's physician.

Still, in order for the therapist to be helpful in establishing a realistic agenda for family therapy, he needs to have a good understanding of what medication can do to help a child and what it cannot do. For example, some medications appear to help many children by decreasing their hyperarousal, their impulsivity, and their hyperactivity for as long as 3 to 8 hours at a time. They may also enhance a child's academic performance by making him or her more attentive in class and helping him or her to complete more work assignments there, with a more legible handwriting. It may even help him or her to reduce negative interactions with teachers and peers.

On the other hand, psychotropic medication cannot help a child to develop motivation or initiative, form lasting friendships, learn to respect and cooperate with others, establish control of his or her emotions and behaviors, or increase his or her intelligence. In other words, medication is of very little help in the reparation of the attachment deficits caused by bonding breaks between caregiver and child.

REFERENCES

Ainsworth, M. D. S., Blehar, M., Walters, E., & Wall, S. (1978). *Patterns of attachment.* Hillsdale, NJ: Erlbaum.

Ainsworth, M. D. S., & Wittig, B. A. (1969). Attachment and the exploratory behavior of one year olds in a strange situation. In B. M. Foss (Ed.), *Determinants of infant behavior* (Vol. 4, pp. 113–136). London: Methuen.

American Psychiatric Association. (1994). *Diagnostic and statistical manual of mental disorders* (4th ed.). Washington, DC: Author.

Armstrong, T. (1995). *The myth of the A.D.D. child.* New York: Dutton.

Barkley, R. A. (1998). *Attention deficit hyperactivity disorder: A handbook for diagnosis and treatment.* New York: Guilford Press.

Begley, S. (1997, Spring/Summer). How to build a baby's brain [Special edition]. *Newsweek,* pp. 28–32.

Belsky, J., Rovine, M., & Taylor, D. (1984). The Pennsylvania infant and family development project III: The origins of individual differences in infant mother attachment: Maternal and infant contributions. *Child Development, 55,* 718–728.

Callahan, J., & Callahan, R. (1996). *Thought field therapy and trauma: Treatment and theory.* Indian Wells, CA: Callahan Techniques.

Castellanos, F. X., Giedd, J. N., Marsh, W. L., Hamburger, S. D., Vaituzis, A. C., Dickstein, D. P., Sarfatti, S. E., Vauss, Y. C., Snell, J. N., Lange, N., Kaysen, D., Krain, A. L., Ritchie, G. F., Rajapakse, J. C., & Rapoport, J. L. (1996). Quantitative brain magnetic resonance imaging in attention-deficit hyperactivity disorder. *Archives of General Psychiatry, 53,* 607–616.

Cline, F. (1979). *Understanding and treating the severely disturbed child.* Evergreen, CO: EC Publications.

Dennison, G. E., & Dennison, P. E. (1994). *Brain gym teacher's edition revised.* Ventura, CA: Edu-Kinesthetics.

Forrest, M. S., & Shapiro, S. (1998). *EMDR: The breakthrough therapy for overcoming anxiety, stress, and trauma.* New York: Basic Books.

Grossmann, K., Grossmann, K. E., Spangler, G., Suess, G., & Unzner, L. (1985). Maternal sensitivity and newborn's orientation responses as related to quality of attachment in northern Germany: Growing points of attachment theory and research. *Monographs of the Society for Research in Child Development, 50*(1–2, Serial No. 209), 233–256.

Haislip, G. R. (1996, December). *Summary report on stimulant use in the treatment of ADHD.* Report presented at the conclusion of the conference of the Drug Enforcement Administration, U.S. Department of Justice, San Antonio, TX.

Halperin, D. A., & Kymissis, P. (Eds.), (1995). *Group therapy with children and adolescents.* Washington, DC: American Psychiatric Press.

Hughes, D. (1997). *Facilitating developmental attachment: The road to emotional recovery and behavioral change in foster and adopted children.* Northvale, NJ: Jason Aronson.

Karen, R. (1994). *Becoming attached.* New York: Warner Books.

Lach, J. (1997, Spring/Summer). Turning on the motor [Special edition]. *Newsweek,* pp. 26–27.

Lahey, B. B., Applegate, B., McBurnett, K., Biederman, J., Greenhill, L., Hynd, G. W., Barklcy, R. A., Newcorn, A., Jensen, P., Richters, J., Garfinkel, B., Kerdyk, L., Frick, P. J., Ollendick, T., Perez, D., Hart, E. L., Waldman, I., & Shaffer, D. (1994). DSM-IV field trials for attention deficit/hyperactivity disorder in children and adolescents. *Journal of the American Academy of Child and Adolescent Psychiatry, 151,* 1673–1685.

LaVaque, T. J., & Rossiter, T. R. (1995, Summer). A comparison of EEG biofeedback and psychostimulants in treating attention deficit/hyperactivity disorders. *Journal of Neurotherapy,* pp. 48–59.

Levy, T., & Orlans, M. (1998). *Attachment, trauma, and healing: Understanding and treating attachment disorder in children and families.* Washington, DC: CWLA Press.

Lyons-Ruth, K., Alpern, L., & Repacholi, B. (1993). Disorganized infant attachment classification and maternal psychosocial problems as predictors of hostile-aggressive behavior in the preschool classroom. *Child Development, 64,* 572–585.

Lyons-Ruth, K., Connell, D., Zoll, D., & Stahl, J. (1987). Infants at social risk: Relations among infant maltreatment, maternal behavior, and infant attachment behavior. *Developmental Psychology, 23,* 223–232.

Lyons-Ruth, K., Repacholi, B., McLeod, S., & Silva, E. Disorganized attachment behavior in infancy: Short-term stability, maternal and infant correlates, and risk-related subtypes. *Development and Psychopathology, 3,* 377–396.

Main, M., & Solomon, J. (1990). Procedures for identifying infants as disorganized/disoriented during the Ainsworth Strange Situation. In M. Greenberg, D. Cicchetti, & E. M. Cummings (Eds.), *Attachment in the preschool years: Theory, research, and intervention* (pp. 121–160). Chicago: University of Chicago Press.

Merrow, J. (Executive Producer). (1995, October 20). *The Merrow Report.* New York & Washington, DC: Public Broadcasting Service.

Milberger, S., Biederman, J., Faraone, S., Chen, L., & Jones, J. (1996). Is maternal smoking during pregnancy a risk factor for attention deficit hyperactivity in children? *American Journal of Psychiatry, 153*(9), 1138–1142.

Nachmias, M., Gunnar, M., Mangelsdorf, S., Parritz, R., & Buss, K. (1996). Behavioral inhibition and stress reactivity: Moderating role of attachment security. *Child Development, 67,* 508–522.

Norton, B. E., & Norton, C. C. (1997). *Reaching children through play therapy: An experiential approach.* Denver, CO: PendletonClay Publishing.

Perry, B. D. (1994). Neurobiological sequelae of childhood trauma. In M. Murberg (Ed.), *Catecholamine function in posttraumatic stress disorder* (pp. 233–255). Washington, DC: American Psychiatric Press.

Perry, B. D. (1995). Incubated in terror: Neurodevelopmental factors in the 'cycle of violence'. In J. D. Osofsky (Ed.), *Children, youth, and violence* (pp. 45–63). New York: Guilford Press.

Perry, B. D., Pollard, R., Blakely, T., Baker, W., & Vigilante, D. (1995). Childhood trauma, the neurobiology of adaptation and 'use-dependent' development of the brain: How "states" become "traits." *Infant Mental Health Journal, 16*(4), 271–291.

Szatmari, P., Saigal, S., & Rosenbaum, P. (1990). Psychiatric disorders at five years among children with birthweights $<$ 1000g: A regional perspective. *Developmental Medicine and Child Neurology, 32,* 954–962.

Terr, L. A. (1991). Childhood traumas: An outline and overview. *American Journal of Psychiatry, 148,* 1–20.

van der Kolk, B. A. (1994). The body keeps the score: Memory and the evolving psychobiology of PTSD. *Harvard Review of Psychiatry, 1,* 253–265.

Wartner, U. G., Grossmann, K., Fremmer-Bombick, E., & Suess, G. (1994). Attachment patterns at age six in south Germany: Predictability from infancy and implications for preschool behavior. *Child Development, 65,* 1014–1027.

BIOGRAPHIES

Alice Eicher Massanari

Alice is a clinical social worker, licensed in both Florida and North Carolina. She is the coauthor of a book about raising a child with special physical needs—*Our Life with Caleb*—and of *Raising Special Kids,* a curriculum for a parenting program on the same subject. Alice worked with Randall Ladnier for several years at a Family Services America agency where much of their theory and research on ADHD was completed. Alice has also been a presenter, trainer, and facilitator for a number of parenting programs in a variety of settings. Currently, she works in an employee assistance program in Asheville, North Carolina while developing a private practice with her husband, Jared Massanari, under the name InterActive Family Resources. They are the parents of two children, one deceased, the other now an independent adult.

Randall D. Ladnier

Randall Ladnier earned his master's degree in clinical social work at Tulane University. He is licensed as a psychotherapist in the state of Florida. Randall has worked as outpatient therapist at Family Counseling Center of Sarasota County for the past 7 years while maintaining a part-time private practice in Sarasota. Prior to that, he worked as a clinical social worker in two privately owned psychiatric hospitals. Randall currently specializes in treating families with recent adoptions and families with children and adolescents who are experiencing problems of emotion and behavior. He has presented at two national conferences of clinicians on the connection between attachment disorder and ADHD. Randall and his wife, Linda, have two teenage sons.

3

PARENTING CHILDREN WITH ATTACHMENT DISORDERS

NANCY L. THOMAS

Therapeutic Parenting Specialist
Glenwood Springs, Colorado

INTRODUCTION

Some of the well-known people with documented attachment breaks who did not get help with healing parenting and attachment therapy are Jeffrey Dahmer, Edgar Allen Poe, Hitler, Ted Bundy, Saddam Hussein, and Ted Kaczynski, the "Unabomber."

> The first clue is something that happened when Kaczynski was only six months old. According to federal investigators, little "Teddy John," as his parents called him, was hospitalized for a severe allergic reaction to a medicine he was taking. He had to be isolated—his parents were unable to see him or hold him for several weeks. After this separation, family members have told the Feds, the baby's personality, once bubbly and vivacious seemed to go "flat" (E. Thomas, 1996, p. 29).

Just as a break in the bond from the parents causes the trauma called Attachment Disorder, bonding with parents can heal it. Parents are a crucial ingredient in the healing of a child with Attachment Disorder. The successful, healing team must consist of empowered, educated parents and a competent, skilled attachment therapist.

Parents of a Reactive Attachment Disordered (RAD) child often silently carry bruises where their child has inflicted pain on their bodies and

Handbook of Attachment Interventions

their hearts, yet they still seek answers and solutions rather than turn away. These child-inflicted wounds are badges of courage. Parents have reported claw marks on their own neck, arms, and faces; hair ripped out; and bite wounds on their arms and breasts from living with a child with RAD. One mother shared the experience of playing baseball with her RAD daughter. She had told her daughter it was not her turn to bat. The mother awoke in the hospital with a lump on her head and was told that her child had knocked her unconscious with the bat. The mother felt sure that it had been an accident. Parents often don't want to believe the worst in their children. They want and need to hope. This 7-year-old girl revealed, in therapy, that it was no accident—she had fully intended to kill her mother because it was not her turn up to bat.

There is hope for these tough kids. These parents need a plan and a powerful team of support to help them work the plan. The parenting plan in this chapter has been developed and honed for over 2 decades by me and others while working with severely emotionally disturbed children. Ninety percent of the children had already killed before they were moved into our therapeutic home where the plan was begun. This plan has proven to be highly successful when completed in the right balance of structure and nurturing. Eighty-five percent of the children placed in my Professional Therapeutic Family using this plan achieved 6 months of respectful, responsible, fun-to-be-with behavior, at home and at school, before being successfully reintegrated into their prior homes. Some of these children had half and many had all of the symptoms of Attachment Disorder. The successful 85% are now graduating from high school and becoming hard-working, loving adults. This plan is most effective for prepubescent children. With teens and adults there is still hope but it is certainly a more difficult climb for them.

One person with a documented break in attachment who did get help is Helen Keller. She became one of our great humanitarians because one person stood strong and did the things necessary for her to succeed. She said, "The most beautiful things in the world cannot be seen or even touched ... they must be felt with the heart." We can make a difference; there is hope for these children.

UNDERSTANDING THE CHILD

In order to understand and have a well-defined perspective of the solutions, we must clearly see the problem. Human infants are born in a helpless and hopeless state. Physically they can neither fight nor flee; emotionally they can do both. When faced with overwhelming pain or loss, such as hospitalized infants or those abandoned, adopted or abused, they will do whatever is necessary to survive. The intense pain and subsequent damage suffered by newborns at the hands of well-meaning medical staff is well documented.

> NICU infants are at risk for being unattached because (1) they have multiple caretakers; (2) the parents are not always present and after a prolonged stay of the baby, may be "strangers" to the baby; (3) the needs of the multiple caretakers and baby may be asynchronous (e.g., it is "care time," but the baby is asleep); (4) lifesaving care in the NICU is intrusive, noxious, and painful; and (5) these experiences give the baby a history independent of his or her parents. From the infant's perspective, the altruistic pain of lifesaving care is indistinguishable from the pain of child abuse (Gardner, Garland, Merenstein, & Lubchenco, 1993, p. 604).

Nancy Newton Verrier writes on the trauma of loss and adoption in her book *The Primal Wound* (1993, p. 1): "When this natural evolution is interrupted by a postnatal separation from the biological mother, the resultant experience of abandonment and loss is indelibly imprinted upon the unconscious minds of these children, causing that which I call the 'primal wound.' " These children are wounded during what Dr. Foster Cline calls the Soul Cycle:

> There is almost a magical cycle that gives birth to our soul. ... The cycle "locks in" our first associational patterns. At any step, if some things go wrong, lasting and severe psychopathology may result. The importance of this cycle cannot be overemphasized. Stage one: Need—Stage two: Rage Reaction—Stage three: Gratification or Relief—Stage four: Trust (Cline, 1979, p. 27).

Because of the trauma these children have endured, they see the world and handle it in a very different way.

> Your child believes that the world is unsafe, that you are unloving, that he is unlovable, that he must control at all costs if he is to survive. Your child believes that you are the enemy and that if you get too close to him, his pain/fear/sorrow will be unbearable (Pickle, 1997b, p. 5).

To help understand the Attachment Disordered child's perspective and help parents to adjust their perception of the child's behavior the following analogy may help:

> Visualize this child in a war zone fighting for his (emotional) life on a horrifying battlefield. Terrified, wounded, and bleeding rather than running into his mother's arms for comfort, the child turns to the earth to dig a fox hole. With the only tools he has, his bare hands, he begins to claw at the earth. Focused solely on survival, heart pounding in fear, tuning out the perceived explosions and whistling bullets overhead, he shuts out the world and digs. Excavating the escape hole for self-protection consumes his every thought, every moment. The battle up above has ended. He does not notice. The dark, cold, lonely foxhole becomes his entire world. Caring people stand near the opening, peering into the darkness he has chosen and offer words of encouragement to come out. He cannot hear them over the racing pulse thundering in his ears and the distance he has put between himself and others. His terror-filled focus is on digging deeper, feeling safer with each foot of earth he flings upwards and out of the pit. The flying debris hits onlookers, pushing many away. Those who care the most stand nearby to call down to the child, bending

> closer to squint into the depth of the chasm in which their child continues to disappear. In standing closer they are often hit full force with barrage of rocks and earth. The child will continue to dig until the walls of the fox hole cave in and they are forever lost or someone offers a better way.

Until the child can see the way out, feel safe enough to try and have someone love them enough to lower a steady ladder, the disturbed child will continue to find ways to remain isolated. As the loving caregivers lower the ladder of skills that require trust and the child grabs hold of the concepts and climbs, each step brings him closer to the top.

The longer they are left to "dig themselves deeper" the higher they must scale to be free and the more difficult the climb. As he reaches solid ground and stands on his own, he is then given different challenges and skills to conquer. As he moves away from the edge onto solid ground more privileges and freedom are given. The more years of childhood that are lost due to attachment difficulties the more opportunities to learn and grow have been missed and the more ingrained the pathological ways of handling close relationships have been set.

PREPARE TO LOWER THE LADDER

The parents' feet must be on solid ground and steadied before they can reach out to their child. They then can lower a ladder with the steps to success necessary for the child to succeed. Parents need to prepare by (1) resting; (2) gaining power through knowledge; (3) gathering a support system; (4) reestablishing authority; and (5) facing the problem.

> Parents must stay calm "in the eye of the storm"; i.e., they stay in control emotionally when the child is out of control emotionally or behaviorally. In order to accomplish this challenging task, parents must always follow the number one rule of good parenting, "Take good care of yourself" (T. M. Levy & Orlans, 1998, p. 199).

Rest for the Weary

Step 1 is both parents must get plenty of rest. The skills and energy required of parents of Reactive Attachment Disordered children are much greater than a parent of a securely attached child. Exhaustion often results from offering opportunities time and time again as a child defies, rejects, and attacks the parents emotionally and/or physically. "Not surprising, perhaps, the most likely candidates for early exhaustion are the parents who are radically committed to their children. After all if there is no 'fire' there can be no burnout" (Dobson, 1987, p. 127). The repeated loss of hope, combined with the "high maintenance" time and skills necessary, create an overtaxed parent no longer able to complete the task at hand. The task, to offer the special opportunities that an Attachment Disorder child needs to heal, must be completed by one primary caregiver. The caregiver, male or female, henceforth in this text will be labeled "Mom." The Mom needs to be rested and prepared for the task at hand. Whether it be lots of sleep or a vacation, both parents must begin the process with step 1, resting.

Knowledge is Power

Step 2 is to acquire more information. Education about this deeply painful condition is a must. Parents often feel tremendous pain having a child that rejects their love.

> 'When parents of unattached children—often children who have suffered early abuse and neglect—are taught child management techniques and then conscientiously carry them out only to see them fail again and again, a sense of futility sets in. Parents then feel increasingly more hopeless' (Cline, 1982, p. 162).

They often feel like "bad parents."

> Because children with attachment disorder don't operate in the world in the same way that securely attached children do, parenting techniques that work well for securely attached children don't work well at all for children with attachment disorder. Normal children (securely attached) can usually be given choices in most areas of their lives, and be allowed to make mistakes because they learn how to handle the consequences of the mistakes they make. However if children with attachment disorder are given choices they will almost inevitably make choices that are extremely detrimental to themselves or to someone else.... And as any parent who has tried to raise a child with attachment disorder can tell you, parenting according to Dr. Spock, Dr. Ginott, Toughlove, Parent Effectiveness Training, Behavior modification, or other approaches to parenting have little to no success with children with attachment disorder either (Randolph, 1997, p. 23).

Books, videos, audio tapes, and seminars focused on Reactive Attachment Disorder may help to empower parents to firm up a plan and prepare to help their child.

All foster parents, adoptive parents, or parents of hospitalized infants should be provided information on Attachment Disorder before they begin dealing with a child at risk due to separation, trauma, or pain. If a child has one or more of the causes of Attachment Disorder, the parents should have a list of the symptoms to watch for and a plan to perform the "healing" parenting required to help these children.

If parents are unable or too exhausted emotionally to perform the parenting techniques presented in this chapter they need to hire Professional Therapeutic Foster Parents. A therapeutic foster parent is highly skilled and trained to work in conjunction with the attachment therapist to treat an emotionally disturbed child in the therapeutic milieu of a family. The therapeutic parents create a healing environment in which to implement the treatment plan on a 24-hour-a-day basis. A professional therapeutic mom is not employed outside the home. In consideration of expertise, specialized training, intense involvement, and high risk factors, these special people are a rare treasure to the families. Many attachment therapists hire these professionals during the intensive beginning of therapy to give the family a rest and begin the tight structure and powerful nurturing required of a child who does not trust.

Children with this disturbance should not be given more information than is absolutely necessary. In the case of securely or anxiously attached children, they cope better with time to assimilate and digest upcoming changes or events. Attachment-Disordered children act worse when given information about what is going to occur. They use it to manipulate their environment and everyone in it. If a child is to have a stressful visit with someone and is told about it they often use the time to obsess, become destructive, and act out their concerns. If something happens and the scheduled visit or event does not occur the child then is angry, disappointed, or sad about it not happening. If information about an upcoming fun event is shared they will often try to sabotage the event beforehand. Sharing information on a need-to-know basis, at the last possible moment, has shown to be the least stressful for all involved and it gives parents the upper hand in the battle for control. When the child asks, "Where are we going?" the parent may ask, "Is there some reason you need to know?" If there is a valid reason then the information should be shared.

The tight structure that is necessary for the healing of these children can be made clearer with a visual picture. The child, having a "war wound," is losing his life blood rapidly through a severed artery. In order to stop the bleeding and save his life, firm pressure must be applied. The pressure bandage must not be moved often to check if the bleeding has stopped. To much blood may be lost with each loosening. It must remain steadily in place until there is no doubt it is no longer needed or the child will not survive. It is painful to hold the pressure for the wounded. They will cry out, "It's too tight, stop! You are hurting me!" The pressure must not be stopped. The parent holding the pressure on the child feels the pain of their child, yet, knows no matter how difficult, it must be done and persists. When the pressure is held steady, long enough and firm enough for the bleeding to stop, no more pressure is required. With Attachment Disordered children the parents must not set up structure out of a need to be in control, military drill sergeant style, but out of love and compassion to save the child they love.

Progress on any level is not possible if the child has not first learned self-control. Do not attempt to have the child handle any chores or privileges until the child has good self-control when mad, sad, glad, or scared. It is vital that they have control of themselves first. The basic tool used to teach self-control is *Power sitting.* Parents select a "think spot" according to visibility, convenience, safety, distractions, and destructibility. A special chair is used for children under 4. A spot on the floor with a small washable, rubber-backed rug for children over 4 is best. Correct body position is with legs folded, hands folded, back straight, head straight, and nothing moving especially the mouth. Having them face a blank wall is safer than having them face a wall covered with wallpaper, which is easily destroyed. Begin with 5 consecutive minutes of power sitting. Build up to 1 minute per year of life. Self-control is crucial if you expect the child to function in a classroom or in life.

Give no negative input. Only positive attention is given for the good parts. Silence is golden; use it to your advantage. If the child is laying down or talking, parents can direct them to lay there or direct them to their room to "rest and get strong enough to try again later." Do not allow privileges until the sitting is correctly completed. Let them take their time doing it wrong first. Time starts when they are in position not when they announce that you may begin timing now. Talking is not part of power sitting.

Sitting is not punishment, it is a thoughtful gift of time for the child to think and get control of him-/herself. There are a number of religions that use this position exclusively to facilitate inner peace, meditation, or prayer. Three times each day for the first 6 months the child needs to have this quiet time to think given to them. Sometimes it will be when they are flashing warning signs and need to think before processing feelings; more often it is used just to practice getting control and getting stronger. Compliance on this basic level is a must before attempting to get compliance, honesty, or any progress on any level. The choice is *X* minutes of "power sitting" or 2 hours of "wimpy sitting" or parents may choose to stand firm on the strong sitting for *X* minutes. The comment, "No problem" means it's no problem for the parent.

The child should also be taught to come when they are called rather than rudely yelling "What?!" from another room. They should go to their room when sent and stay there. These are all things that can be practiced when it is convenient for the parents.

Gather A Support Team

Step 3 is to gather a powerful team together to support the parents' work. Just as elephants form a very tight circle join their bodies together side by side, facing out, to assure the safety of the encircled baby elephant, an awesome team needs to be joined together for the RAD child. The circle of support may include the parents, therapist, grandparents, the teachers, social worker, friends, respite care provider, and the church community. Those interested people who can be strong enough not to be manipulated by the child should be invited to the team. The support team needs to have a clear understanding of Attachment Disorder and the plan. Tremendous things can happen when all of the people involved in this child's life link together in a powerful human chain to stand and do what is right for this child. The strong parenting required to stand firm and set the limits that need to be set is very difficult without a tremendous amount of support. A single goose can fly a long distance, their powerful wings can lift them high and fly far, but in a flock, as a team flying in formation, they can fly 71% farther (Canfield & Hansen, 1995, p. 307). The parents are the ones who begin the circle. They need to be a powerful united team. Disturbed children need a parenting team. It is extremely difficult for a single parent to be the mom and the dad and the financial support system that is necessary for a RAD child to heal. People need to team up so that there is a united front to handle the difficulties that this child is going to throw at them. When one parent is tired, the other parent can fill in. One draft horse alone is able to pull 2 tons. Two draft horses sharing the load as a team can move 23 tons (Canfield & Hansen, 1998, p. 305). We are without a doubt much more powerful when we are united. Two single parents or adult siblings can join together as a team to parent these kids. It has been done very successfully.

The people in the circle need to be nonjudgmental. The team should not be looking to place blame, but solely supporting and seeking solutions to the problems. "I believe that the impact of the child's trauma upon the family system is greatly underestimated by clinicians and that the focus of the dynamics is skewed to seem as if the problem resides in the parents' issues" (Verrier, 1993, p. 2). Donovan and McIntyre (1990; cited in Verrier, 1993, p. 3) pointed out that their findings demonstrated a "striking consistency of behavior problems among adoptees whether the family is functional or dysfunctional." Judging people without the information required or even with it keeps us from being supportive and helpful. It can be and is very damaging:

> A profound learning experience I had about judging folks occurred on the Fourth of July. I had been ruminating about the importance of not condemning parents for mistakes. We had a big gathering with hundreds of families on the grass in the park awaiting the fireworks. One child that I could not see was squealing like a toddler. A few of those high-pitched screams was okay, then it continued on. I quickly decided it was obviously inept parents that needed to teach their child the concept of "no." There was someone rows behind us who began to yell, "Shut that kid up!" I shared my thoughts with my daughter, who looked at me in surprise and said, "Mom, that's a severely handicapped little boy who can't help it." Ashamed, I realized I had judged the parents without all of the facts. I had condemned them.

Our society often reacts that way. To be successful in supporting each other we have to stop judging and laying blame. Even therapists have been guilty of this: "At Bellevue Harry Bakwin had the habit of telling distressed mothers that 'there are no behavior-disturbed children, just behavior disturbed parents'" (Karen, 1994, p. 27).

Our eyes speak reams, volumes of information about whether we believe in that parent or whether we don't. We need to be very aware of that. We need to support with our eyes. We need to look at parents with honor and love that they brought this child into their home, into their lives, whether through birth, adoption, or foster care and are doing their best. Stephen Covey (1997) in his book *Seven Habits of Highly Effective Families* talks about the family as a plane headed for a destination moved off course by winds and various encounters. We are all off and on course continually. It is part of being human. Keeping the goal in mind helps us get where we need to go. The support system's goal must be to help parents stay focused and encouraged.

Relief care is an essential piece in the support team for the family. The relief care provider must be trained and selected carefully. This provider must maintain the essential tight structure and must do no bonding activities with the child such as eye contact, touch, and smiles, and sharing sugared foods. The relief provider should support the parents with their difficult work with the child.

Establish Authority

Step 4 is to reestablish authority in the home, especially with the kitchen, the car, and the cash. A child who is "out of control" needs someone to control them until they can build *self-control.*

Food that is prepared for children is a way of nurturing. They will often reject homecooking, choosing self-prepared or fast food to avoid accepting love and human connections. The most successful method to deal with this is for the parent to prepare and provide all meals. There being no other available options, eventually the child will join the family for meals. Rather than forcing food at the child it works better to do the reverse. If a child makes a face or rude comment about a meal that has been lovingly prepared, the plate is quickly and calmly removed. It can be thrown out or, better yet, fed to the dog with the comment "Don't worry, the dog loves this food!" When the child asks, "What am I going to eat?" the loving parent may say, "Breakfast!"

The car is only used to transport children that are "climbing", not digging deeper with behaviors to push parents away (unless it is an emergency or a trip to therapy). The child should sit in the back seat behind the passenger seat until they have completed climbing the ladder of success (unless there is an airbag safety issue). Many powerful parents do not allow disturbed children to speak unless spoken to in the car for safety reasons. Having the child ride with a hand over the weak mouth for 5 to 15 minutes has helped children learn not to distract the driver. If the child has weak hands and is hitting other children or playing with windows they may need to ride for 10 to 15 minutes with both hands on top of their head until they get control of their hands. Sometimes practice sessions are required when the driver is less distracted and has returned home. These practice sessions can be done in a mock car seat (chair) with hand over mouth or hands on head for twice the time originally asked. The child should take their time and not rush into putting their hand over their mouth until they are ready. Patient parents wait calmly by enjoying the time reading a good book or playing with other children. They consequence the first word or flying hand rather than wait until they are angry.

A driver's license is not a wise option for a child out of control. They must be demonstrating excellent self-control over an extended period of time before they (and the rest of us) are safe on the road. Money is another area disturbed children have trouble handling appropriately. The parents should keep and control all cash until the child displays the skills to make good decisions concerning their life. Drugs, cigarettes, alcohol, lies about thefts, and bus fare for running away are all made easier with cash in hand. Retain control of the kitchen, cash, and car until the child has climbed from the hole they have dug and are on solid ground for at least 6 months.

Face the Problem

Step 5 is that adults involved with a RAD child must have their eyes open. It is highly likely that the child will attempt to manipulate because they do not trust that others have their best interest in mind. Parental denial can prevent a child from feeling safe enough to trust adults. Each time parents close their eyes, in denial to the child's behavior, it is equal to wobbling that ladder. The child believes if he/she can outsmart and manipulate the adults, no one is strong enough to protect them. The banners of belligerence that these children carry of lying, stealing, breaking things, fit-throwing, sexually acting-out, hurting animals, and oppositional defiant behavior are thundering cries for help. Children will fail when parents turn a deaf ear or warn, lecture, or explain rather than have the child face the consequences and make restitution. For example, one mother came out to check on her beautiful puppies only to find the heads had been ripped off each one and lined up in a row. Horrified, she decided to never breed the dog again, assuming it had been the dog who had killed them. It was later discovered that they were killed by her 5-year-old adopted son. Parents often do not want to assume that it is their child's doing when the dog is limping or the cat is missing. The child that killed those puppies then went on to kill more animals, as he was crying out for help with his actions rather than his words. If parents close their eyes to the child's cries for help, they become as blind as the child; the child is blind to their love. When a parent finds drugs in their child's room, they must not close their eyes and ignore this desperate cry for help. Appropriate action must be taken promptly. When stolen property is discovered, it must be actively dealt with. These children do not learn from words; they learn from action. To recover from the problem of stealing they must make restitution for every stolen item and every suspected stolen item. They need to make restitution for injuries inflicted. They earn the money and pay it back or they give of their time to lift the load of the one they insulted or hurt. One option for restitution is to do chores (e.g., vacuum the bedroom, fold the laundry for siblings). They can rub lotion on the feet of the tired Mom they have hassled (this becomes a favorite of many moms).

SHINE A LIGHT ON THE SUBJECT

A light must shine to illuminate the child's path into the parents' arms. The most powerful light source is the parents' eyes. Filled with hope and love, shining onto the child, they can be a powerful beacon beckoning the child toward emotional stability. It can be the light of hope to illuminate the cavern they have dug themselves into.

The sound of laughter is a sound that says "all is right with the world" and it is a safe place to be. King Solomon said, "Laughter is good medicine." Laughter is the shortest distance between two people. Studies have shown that after we laugh we go into a relaxed state. Laughter lowers the heart rate and blood pressure. It also indirectly stimulates the release of endorphins, the brain's natural pain killers. Laughter increases creativity and problem-solving. Norman Cousins, in his 1979 bestseller *Anatomy of an Illness,* describes how he cured his own fatal disease with laughter. Studies have shown how it activates killer and T cells as well as increases immunity to disease (Doskkoch, 1996, p. 1G). Laughter is, of course, never at the child's expense. Laughing at the child puts you against, rather than for, the child. In this case, they will not feel safe enough to turn toward you and trust that your outstretched hand will hold them. The sound of laughter has been clocked at over 700 MPH. Kindergartners laugh an average of 300 times a day and an average adult laughs 17 times a day. In the home of an attachment-disordered child there is little laughter. Bring laughter back into the family as a healing tool for all. Set it up and plan it. Eat pizza in the bathtub. Paint on a mustache and fix breakfast. Dance a jig. He who laughs, lasts.

It is said, "Music soothes the savage breast." Appropriate music can help calm a disturbed child. Many have found it helpful. Mozart's music has proven to have a calming, soothing effect that opens the mind to be more creative and have better problem-solving ability and increased memory. Mozart's melodies can substitute for sedatives, boost memory and concentration, and improve thinking (Campbell, 1996). It actually improved students SAT scores by 51 points on the verbal portion and 39 points on the math portion. The result of listening is a nurtured feeling that supports well-being. It is not a cure, but, it can be helpful in the car, school, therapist's office, or at home.

The smells of love wafting toward the child—of homebaked bread, of a good meal in the oven, of herbs used in aromatherapy (chamomile, lavender and lemon balm for calming and stress relief)—can have a helpful effect on healing the heart of a child and the entire family. In research lead by Robert Baron, a professor at the Rensselaer Polytechnic Institute in Troy, New York, they found that with a pleasant fragrance of roasting coffee or baking cookies in the air people were more than twice as likely to be helpful and pleasant than in unscented surroundings. Baron said, "The effects of pleasant fragrance on social behavior stem at least in part from fragrance induced increments in positive affect" (1996, cited in Crenson 1996, p. 4A).

The feeling of peace that emanates from a parent who is organized and in control demonstrates strength that a disturbed child must feel in order to develop trust. Having the household organized with a chore list so that dishes, bill paying, checkbook balancing, laundry, and so on can be done on time takes some of the stress off the family with a high-maintenance child. This allows more quality time to assist the child.

The presence of a full-time parent for the first 6 months to 1 year of the child's healing time is highly recommended. The act of prioritizing and committing to the child's needs demonstrates the parents' commitment to the child's emotional survival. If that is not possible, a primary caregiver, such as a day care provider, must be enlisted to perform the necessary bonding and skill-building. This person must be selected carefully to replace the parents in the bonding and guidance of the child: "... the parents need to recognize that the alternate attachment figure may now count more prominantly in the child's development" (Karen, 1994, p. 336). "A study of 110 families by Peter Barglow found a shocking correlation. When full-time nanny care began in the first eight months, it was associated with a much higher incidence of insecure attachment to the mother at the end of the first year" (Karen, 1994, p. 341). Barglow's study was done under optimal conditions with all the parents being "well educated" and deeply involved in "parenting." The day care provider should be one that will remain in the child's life for a minimum period of 2 years. At the time of departure the bond must be carefully and slowly transferred to the parent or another caregiver. Each move puts the child at a higher risk for severe emotional problems. A bond cannot be quickly severed without retraumatizing children and forcing them to return to digging (old anxious or avoidant behaviors) in terror for their life. In a presentation for the Attachment Center at Evergreen in 1994, Greg Keck, Ph.D., stated emphatically, "The term 'moving a child' should be called 'inducing loss' in the child." Children should not be moved from caregiver to caregiver, especially during the first 3 years of life or during the healing of previous attachment trauma.

REMOVE OBSTACLES AND BARRIERS

One of the most difficult types of child to treat is one that is unattached and overindulged. To evaluate if a child is overindulged, assess their balance of give and take. Are they under 18 and have a double bed, TV in their room, private phone, large shelves filled with toys, an extensive wardrobe, and/or more than one bike? Do the parents rescue them from consequences the child needs to handle? Does the child eat what is served or is the entire family fed according to the child's food choices? Does the family serve the child, while the child does little or nothing to reciprocate?

Overburdening a child with excessive material goods and privileges is akin to having him or her climb the ladder from the dark depths of the fox hole carrying all he owns. The handicap of climbing with only one arm and of having the view of the parents obscured, as the "stuff" blocks the light from above, would make it extremely difficult, if not impossible, for the child to succeed.

To prepare the child for the ascent it is essential to remove obstacles and teach self-control. The child's room should be prepared calmly, while they are elsewhere, with: a single bed, bedding, blankets and pillow, a maximum of 7–10 sets of seasonal clothing, socks and underwear, two coats, three pairs of shoes/boots, one hat, and a few easy-to-pick-up, age-appropriate toys and books. A desk with supplies that the children can handle alone should be provided for those that attend school. Legos, art supplies, books, and a mini-trampoline should be made available in the family living area. In order for the child to focus on the goal of connecting to the parents, their room must be easy to organize and maintain. Keep it simple. Store or dispose of the rest.

Reactive Attachment-Disordered children usually have trouble with empathy for people and pets. In the beginning, to help them stay focused on the goal of learning to be respectful, responsible, and fun to be around, they should be prevented from interacting with other children and animals. After several weeks of being successful with the goals, these privileges can and should be earned. If the child has a problem with sexual perpetration or predatory behavior then 6 months to a year of success with the goals required would be more appropriate.

The child's home should be prepared by removing items that are treasured by other family members. This would include heirlooms, jewelry, credit cards, and any such items that would cause distress to the family if they were lost or destroyed. The child must be a priority over material possessions.

> Meaning doesn't lie in things. Meaning lies in us. When we attach value to things that aren't love—the money, the car, the house, the prestige—we are loving things that can't love us back. We are searching for meaning in the meaningless. Money, of itself, means nothing. Material things, of themselves, mean nothing. It's not that they're bad. It's that they're nothing (Williamson, 1992, p. 16).

STEADY THE LADDER

Parents need to hold steady by using powerful parenting techniques that do not include anger, warnings, second chances, or waiting to consequence until the parent has become stressed. Anger feeds the RAD child's pathology and assures them they are right in not trusting. The flinging dirt from their digging often causes parents to think the child is "out to get them" rather than to see this as a side-effect of the RAD child's problem. Warnings ("If you do that, I'll have to ...") and second chances ("If you do that one more time ... The next time you do that I'll ...") are not effective with these children and should not be used. These behaviors keep a child from trusting that the parents are strong enough to keep them safe. It would be similar to lowering a ladder for the child to climb up and making it wobble as they reach for it. The ladder must remain steady. Consequence the *first time* the child does inappropriate behavior or crosses set boundaries. Use action not anger.

Another major obstacle to the child's progress and subsequent healing is the holding of destructive secrets. These would be secrets concerning lies, stolen goods, injuries to others, and any inappropriate behaviors. A complete "clean slate list" is often required of the child before they may begin to earn any new privileges. The therapist usually assigns this task to the child during the beginning phases of therapy. Close, constant supervision to prevent old destructive behaviors, such as molesting children or hurting pets, must be provided until the child develops adequate self-control. Supervision throughout the night requires the use of an inexpensive intruder alarm on the child's bedroom door and window. When installed the child should be told the alarm is there to keep them safe from anyone entering to harm them and from them coming out to get themselves in trouble. The alarm must remain in place every night until the child has been on solid, stable ground for one year or more, depending on their level of conscience development.

Yet another obstacle to healing and connecting can be the acceptance or rejection of the child's past.

> Parents need to be non-judgmental and accepting of the child's previous life. It is part of the child. To reject the child's perception of his previous life experience is to reject the child. The more we can share the child's past through their eyes the more we can walk supportively and acceptingly with the child. Lifebooks help to

make sense of a child's history. They give tangible evidence to that experience. They provide connection and understanding to a child (Pickle, 1997a, p. 10).

Rejection of the people who gave the child life is extremely damaging to the child. They are, and always will be, a genetic part of them. Pictures of the people and places in the child's life should be gathered and put all together in *one* book to give continuity to their life.

Sleep is a crucial ingredient to learning, growth, and sanity. Researchers at the Weizmann Institute, in Rehovot, Israel, and at the University of Arizona found that a good night's sleep is an essential ingredient to memory and learning (cited in Reuter, 1994, p. 6c). Children need to be put in their rooms in a regular nightly routine; younger children at 7:00 P.M., older children at 8:00 P.M. This gives parents a break from the full-time job of parenting and gives children a quiet time to do homework, read, relax, or play quietly before they put themselves into bed. Attachment-disordered children with trust issues often insist on having a night light and the door left open as a control attempt. This should not be an option for a child over 5. If a child has trouble sleeping, it needs to be remedied quickly. Sometimes warm milk, a back rub, or some herb teas are helpful. Sometimes medication is required. Lots of exercise has shown to make a big difference in the soundness of sleep. The use of the mini-trampoline daily makes an impact. It is vital that the child gets 8 to 10 hours of sleep every night.

TURN THE CHILD TOWARD THE PLAN (THE LADDER OF SUCCESS)

To turn the child from their fight or flight toward the ladder of success in life there are two requirements: the first is Attachment Therapy, the second is removal of distractions.

Attachment Therapy

Attachment therapy is required to dissipate enough of the internalized rage to make room for love to enter. It is essential to find an effective therapist. There are no excuses for the professional community not knowing about Attachment Disorder. It has been over 50 years since the United Nations decided to commission a study of the problems and needs of homeless children (Karen, 1994). The World Health Organization commissioned John Bowlby to investigate and report his findings. In 1951 his findings, published as *Maternal Care and Mental Health,* changed the way the mother–child connection was viewed. What emerged from this inquiry, reported by Fraiberg (1977), was that even the life-threatening dangers of war were not as destructive to the minds and emotions of children as separation from their mothers and fathers.

> Bowlby argued that the mother–infant relationship is an extremely important one, that it was not a pleasant amenity for the child but an absolute necessity and that significant early separations are perilous to the child and ultimately to society as well.... He advocated that large numbers of people be trained in marriage and child guidance and in work with parents of the very young. He said the large outlays of funds required would be far less than the later costs of institutional care and delinquency (Karen, 1994, p. 62).

In 1980 the *Diagnostic and Statistical Manual of the American Psychiatric Association* (DSM III) added the term "reactive" to what Foster Cline, M.D., had termed "Attachment Disorder" in 1972, when he co-founded the Attachment Center at Evergreen. This information has been available for over 5 decades.

Misdiagnosing these children, when they have documented one or more of the risk factors for attachment problems, causes the families to waste financial resources and, worse yet, precious time in inappropriate therapy. This is inexcusable. Some of the children who are shooting their classmates and killing teachers have been misdiagnosed and are using the wrong type of therapy. Kipland Kinkel had been in traditional therapy and on medication for years before he killed his parents, prepared a meal and ate it with their bodies present, and then shot and killed his classmates and teacher. He clearly had no conscience. Effective help for RAD children must include attachment therapy. To turn the child from their focus of fight or flight, they need to begin to trust the world as a safe place. They need to see the adults in the world not as the ones who left them or hurt them but as those who are capable of helping and being trusted. This is the purpose of attachment therapy. The therapists job is not to focus the child on the therapist but on the parents and to turn the child from repetitive digging to facing the ladder of help offered by the parents. The RAD child views the mother as the main target because, to an infant, the mother's job is to keep them safe. As a baby they were not kept safe. "It has been shown that regardless of the intellectual reasons a child has been given for his relinquishment, there are often feelings of betrayal, anger, resentment and sadness, which are projected onto the available mother-figure" (Verrier, 1993, p. 56). The Mom is blamed and targeted. The therapist must use the mother as the "change agent." Daniel Hughes, a psychologist specializing in attachment therapy in Maine, says, "With the parent present in therapy there are numerous ways to communicate to her child the value and importance of parental authority. The therapist shows the child how she respects his parent's judgment and skills and how she defers to the parent in decisions about the child" (Hughes, 1997, p. 80). "With the help of a competent therapist, parents can change the way their problem children act and create a future full of hope for their children and themselves" (Cline, 1995, p. 161).

There are skilled attachment therapists in 32 states in the U.S. listed in 1995 in *Give Them Roots and Let Them Fly* published by the Attachment Center At Evergreen, Inc. (McKelvey, 1995). Since that writing at least five more states, as well as other countries, have been added to that list.

Attachment therapy must be the precursor to other therapies such as play therapy, sand tray therapy, art therapy, talk therapy, EMDR, or any other type that requires a therapist/client relationship and honesty. One textbook, referring to the level of rapport required for this therapy to be effective said,

> Clients should be able to feel comfortable with the possibility of experiencing a high level of vulnerability, a lack of control and any physical sensations from the event that may be inherent in the target memory. This means that clients must be willing to tell their therapist the truth about what they are experiencing (Shapiro, 1995, p. 91).

These children do not have the ability to form relationships or to trust others enough to be honest about their feelings. They are not ready to give up control without a fight. That is the nature of the beast we call Attachment Disorder. In one of his studies David Levy (1943; cited in Karen, 1994, p. 16) found that none of the children [with at least 10 symptoms of RAD] seemed able to respond to psychotherapy. Only after the child heals that part of their damaged psyche with attachment therapy can these other modes of therapy be used effectively.

Remove Distractions

To help focus the child on the goal and motivate them to reach for it, we tighten the child's world into a smaller sphere of activities and remove distractions. The recommended beginning activities are building with legos (or other similar construction type toys), art projects (coloring, drawing, painting), reading, and exercising on a mini-trampoline. All of these activities can be enjoyed within the parents vision. Just as an infant needs to be held close and guarded carefully, so does a child with little or no self-control:

> It is important to understand the rationale for this structure and control. The younger the child or the more disturbed the child, the more structure and control a parent must put into place to provide maximum protection for the child, Children with Attachment Disorder tend to be self-parenting and rejecting of parental authority as well as parental nurturing. In order to parent these children, we must structure the environment in such a way that these children begin to see parents as capable of providing safe structure and providing nurturance (Pickle, 1997a, p. 23).

Television and movies are a major distraction to the child's healing that must be completely eliminated. Statistics from TV-Free America, a Washington, DC based group led by Henry Labalme, say:

> The average 18 year old will have seen 200,000 violent acts and 10,000 murders in their homes via the tube. Sixty-six percent of Americans watch the tube rather than converse with each other during the important time of sharing family meals. 1,680 minutes each week the average child watches television. 38.5 minutes a week the average parent spends talking to their children. 9 years of their lifetime the average viewer wastes invested in TV during a 65 year life (TV-Free America; cited in Briggs, 1997, p. 1E).

Prime-time needs to be redefined so that it means time with the parents not the tube. Pope John Paul II said, "Parents who make regular, prolonged use of television as a kind of baby-sitter surrender their role as the primary educators of their children" (cited in *Rocky Mountain News,* 2/25/94, p. 54A). Time must be invested in working on their lives. Many of these children have already lost years of their childhoods to the seething rage of this disorder; they cannot waste any more of it on something that hurts more than helps. Attachment-Disordered children need to be unplugged from the TV so they can plug into life. After the child has healed, 2 to 3 hours of viewing per week may be earned as a special privilege.

Children need to play every day. Playtime can be used as an excellent motivator for the child to start accepting limits. Time for lego-building may be given with the guidelines of playing in a specific 4 ft. × 5 ft. area with no talking. The child is then allowed to play with the Legos until one toe or one toy crosses the set boundary or one sound is spoken. The parent then helps the child clean up the toys. The art supplies are brought forth with two limits—stay in the chair and create quietly. When a limit is breached, the parent calmly helps the child clean up; no discussion, no lecture, no argument. A book is selected by the parent and given with two limits and so on. In the normal course of the day, the child must certainly be allowed to speak but during these brief training periods the child must practice skills for working in school where remaining at one's desk quietly is essential.

THE STEPS TOWARD SUCCESS (THE LADDER)

The two legs of the ladder are Parenting and Attachment Therapy. Both of these must be in place for the plan to hold. The eight rungs of the ladder that support the child while they ascend from their pit toward the light of love and success in life are listed below.

- Respect for people, animals, and property of others
- Responsibility for their body, their possessions, and their job
- Fun to be with by displaying a good attitude and appreciation

Respect for People

Bowlby's definition of attachment is "an affectional tie with some other differentiated and preferred individual who is usually conceived as stronger

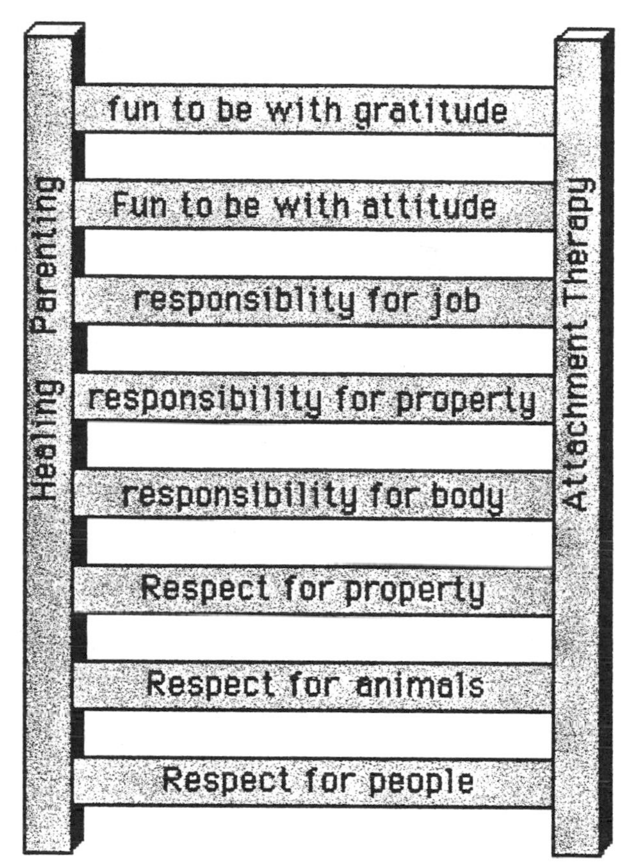

FIGURE 1. The ladder of success.

and/or wiser" (Bowlby, 1977, p. 203). Commanding respect is how parents can begin to be that person who is conceived as stronger and/or wiser.

> Parents, teachers and other adults who tolerate disrespect are saying to children, "I am not worthy of respect." The children then say to themselves, "If I can treat others with disrespect and get away with it then people can treat me with disrespect and get away with it. None of us are worth anything at all, not even me." The damage done to a child's self-esteem when he or she is allowed to whine, curse, swear and, in numerous other forms of speech and behavior, be disrespectful of the loving authorities in their life is incalculable (Hage, 1997, p. 5).

Respect for people begins with making and maintaining eye contact, especially when speaking or listening to parents. What either parent or child has to say deserves full attention: "Eye contact" is not just looking at the child's eyes. It is reaching into their soul with eyes that are rested, loving and powerful. "Eyes that say to the child, 'you're okay, you have me.'"

Parents can reinforce this message to the child over and over, throughout the day throughout life. Eye contact is POWERFUL—even brief eye contact. It can be an expression of love as well as a weapon of pain. You must be very conscious of how you are using this powerful tool. Keep it soft and loving. One brief look of daggers from parents' eyes can undo weeks of work (N. Thomas, 1997, p. 32).

In the beginning parents must encourage the child to make eye contact even if it is just a brief moment. It is recommended to discontinue conversation when eye contact is broken. Wait quietly to see the eyes again, then continue speaking or listening.

The child should be respectfully opening the door for parents. The child should learn to be respectful by walking beside or behind parents rather than pulling ahead or lagging behind. Walking in a thoughtful way, honoring parents as their leader, makes a clear statement of respect.

To ingrain the positive behavior, it must be noticed and "pizzazzed." Parents can look in their eyes, smile, and with excitement say, "Yes! Good job with the eye contact! You are getting stronger!" In the beginning, children with RAD do not do things to make parents happy until their hearts are mended. They actually prefer negative attention to positive. In the child's mind making someone else happy means that they are "losing." They want to be strong because they do not feel safe. Strength and power are what these children seek. Parents can use that insight to help them heal. "I see your eye contact is getting stronger. You're strong enough to look me right in the eye! Good job! That's it! You're getting better!" Those terms are needed continually throughout the day, with enthusiasm. In the beginning "pizzazz" is used for many of the little things. Later, it is used mainly for more important behaviors. "All right, you made it home on time! Good job. Look at you. You are really acting responsible. You are really starting to make some good strong decisions here." That positive feedback from a loving parent is what helps to shape the child into a successful adult.

The best time to "pizzazz" is right when the behavior happens, but if the opportunity is missed, it is still effective to say, "You know, earlier today (or earlier this week), when you opened the door for me, I was impressed that your brain was so strong! You were thinking ahead when you remembered to treat me with respect. Good for you! You are getting stronger!" Rewarding that behavior with positive encouragement is essential to get the child to repeat the behavior. The behavior receiving the most pizzazz is the one the child will repeat.

Eventually they can learn to be respectful and responsible because they will feel good about themselves. Similar to toilet-training a toddler, every time they "go" correctly, enthusiastic praise is offered. When was the last time anybody appreciated you for going in the toilet and not in your pants? Adults continue for years doing appropriate behavior with no more positive reward because they feel better about themselves and it is a lot easier. It is the same for these children.

Occasionally these children have a depletion of oxygen to their brain which causes them to stare at parents and answer or say "What?" or "I don't know" instead of answering respectfully. When these things occur a skilled parent will require the child to do push-ups or jumping-jacks. The number of push-ups or jumping-jacks is decided according to two push-ups or jumping-jacks per year of age of the child. These children require a great deal of exercise. A loving parent looks for opportunities to provide it with love.

Some parents get confused as to what their job is. The job of a parent is to offer opportunities for growth to their child. The child uses or wastes these. It is not the parent's job to change the child.

> A child who knows his problems are the concern of another, concerns himself with none of his problems! ... Tell a fine lazy person that he's basically lazy and he'll love it while standing in line for food stamps. On the other hand, take away the food stamps and make finding the meal his problem and suddenly there is a definite satisfying grinding shift as the old rear is put in gear! (Cline, 1982, p. 6).

The parents' job is to offer the love. The child's job is to take it.

Respecting parents' need for quiet and other people's time for talking is another important area of respect. Emotionally healthy people use their words to convey thoughts and feelings and to ask questions to gather information. Attachment-Disordered children use their words to control, interrupt, and make noise. These behaviors are not appropriate and must be eliminated. Having them put their hand over their mouth when "their jaw gets weak," so weak that they are not able to keep it closed without extra help, teaches some self-control. Once they get their hand up there then use positive rewards by smiling and saying, "Good job getting your hand up." That positive feed back has to be there. Then after their hand is up and they are quiet for 10 or 15 seconds parents can then say, "Good, you are getting stronger now." After about 30 seconds say, "You may take your hand down now. Let's see if your jaw is strong enough to keep your mouth quiet. Good job! I don't hear any noise falling out. That is so excellent! I see you're doing a real good job of getting your jaw stronger by keeping your mouth closed and quiet. Good self-control."

When they do speak, it needs to be very respectful. They should be saying, "Yes Mom!" "Yes Dad!" "Thank you" and "May I please have a drink of water?" Their communication needs to be complete, clear, concise, and respectful. It should not include shoulder shrugs as communication or grunting noises such as "uh huh" and "huh?" Those are not respectful ways to communicate. "Yes Mom," "Yes Dad," using the parents' title of honor, each time reconfirms in the child that title of authority.

Asking for things that they need, such as a drink or to use the bathroom, is part of replaying that first year of life when their needs were not met by the primary caregiver. In the beginning, they must ask for everything. Complaining as a way to get their needs met is not appropriate. They should not be allowed to communicate their needs by griping, such as "I'm thirsty" or "I have to go to the bathroom really bad." That is not a respectful way to get needs met. They often try to get their needs met by disrespectfully commanding and demanding. Sometimes they even put a please on the end such as "Give me a drink, please." A child must learn to respect and to look up to an adult enough to trust them in order to heal. Benjamin Franklin said, "Let the child's first lesson be obedience, and the second will be what thou wilt."

Another problem of disrespect that children with Attachment Disorder sometimes have is whining. The first time a whine noise comes from a child, they are immediately put in their room for a nap for 30 minutes. A good parent always listens to their child and when a baby starts getting tired and cranky they whine to announce that they need a nap. Parents can tell the child, "Please whine to let me know you need a nap." Threatening, "The next time you do that you are going to take a nap," is completely ineffective. The instant they whine they are given time for a nap. Expect resistance when they are put in their room to rest. The 30 minutes starts when it gets quiet. Family members must be silent while the defiant one goes through the tantrum of seeking negative attention or any attention for negative behavior (kicking walls or the door, ripping off sheets, curtains, etc.). Absolutely not a word from the family is spoken until it is quiet for several minutes. That is then the time for an encouraging "I hear you are getting strong. Nice job being quiet. I will start the time now!" No one should have a conversation or argument with the "resting" child. If the child tries to start one, walk away in silence.

Arguing, another disrespectful behavior, takes two. Have them put their hand over their mouth the first time they begin an argument. Parents must not "pick up the gauntlet" they throw down by responding.

Have them practice handling the word, "No." Practicing parents can say, "I want you to think of five things to ask me that you want to have so you can practice handling 'No.'" After each thing that they ask, say "No." By the third one have parents say, "Oh you've just handled me telling you 'No' three times without whining, fussing, or arguing. Good for you! Good job! I see you are getting stronger!" Then have them ask the other two things and each time say "No!" and reinforce their quiet acceptance. Reaffirm in them that they can handle the word "No" in a respectful manner.

Using the term "why?" is yet another disrespectful habit often seen. These children are not asking questions to gather information such as "Why is the sky blue?" they are asking "Why do you think you have the authority to tell me what to do?" A good reply is, "That is a really good question. When you're finished doing what was asked, I want you to write a paper answering that question. When you have a full page, bring it to me and we'll talk." If they are too young to write then they should be told to think about it for a time and then to explain to the parents.

Lying is one of the most frequently encountered problems families deal with. When a child lies, it needs to be dealt with. First, parents must not believe them. The more adamant they become that they "didn't do it, didn't do it," the higher the odds that they *did* do it. Assume that they are guilty and if it is later learned that they are innocent, then, of course, return what they have paid for in damages and apologize. Restitution for lying is very important. They can pay back with some of their time by doing an extra chore to help the family, rubbing lotion on parents' feet, or anything where they must give of themselves. They do not get negative pizzazz for their lies. "Don't you lie to me!" is not effective. Do not get emotionally upset. Calmly say, "Gosh I'm sorry you were so weak there. That lie just fell right out of your mouth. I hope that you'll be able to get stronger. But in the meantime what you can do to make up for wasting my time with your lie is.... " A different way of handling lying is to purchase a special notebook. Explain briefly to the child that it will be interesting to see how long it takes him to get strong enough to tell the truth after they lie. When the child lies have the parent smile and rush to write it in the notebook. No discussion about what the real truth is. When the child does tell the truth the parent should rush with glee to the book and document it with enthusiasm. The score card will show the child telling the truth sooner and sooner until eventually the lie will immediately be followed by the truth at which point the liar will be strong enough to soon stop completely.

It is respectful to listen to what people have to say the first time. RAD children often say, "What?" This is a little test that they use when meeting a new adult or to exhaust an adult they are trying to control. Within the first few minutes of meeting an adult, the adult will say something and they will reply, "What?" to see if they can get the adult to repeat themselves. This is the time to pass, rather than to flunk, the test. The wise adult should say, "What was it I just said?" The child will often repeat word for word what was said. Sometimes they say "I really don't know what you said." Then say, "Well, my time is very valuable, I need to make sure that your brain is working before I go to the trouble of repeating myself. So, I want you to do ten or twenty-five jumping-jacks [depending on how old the child is] to get the blood supply moving to your brain to carry in some oxygen." After compliance is gained and the push ups or jumping jacks are successfully completed, repeat the information. This way the child wins by getting the information repeated and the adult wins because compliance was given. When both win the battle, the child can feel safe enough to succeed at learning to trust.

Interruptions by the child can drain parents' energy and often convince the child they are in control. One solution is to have them put their hand over their mouth to work on developing strength. In order to be most effective it should be done at the first word of interruption. We have two ears and one mouth. They need to be used in that proportion. Parents need to listen twice as much as they speak. Children need to listen twice as much as they speak.

Name-calling and foul language can be problems of disrespect. When these children are calling people names it is often a projection of how they feel or of areas they are concerned about. Rather than getting angry about the term, it is more beneficial to focus on the information the child may be attempting to convey. The consequence for foul language that has been shown to be successful is to have them scrub toilets, shovel manure, or scoop poop. Filthy language problems (or having "manure" come out of their mouth) can be solved by having the child develop some strength while moving manure. Most livestock are not bothered by foul language.

Children need to learn to accept, respectfully, a limit being set. Some whine, some throw fits, some argue in an attempt to regain control. A RAD child will often go into an infantile regression, laying on the floor with their entire body wracked with screams from deep within. If the parents can remain patient, calm, and are physically able, they may prefer to hold their child in their arms while the child rages. This method demonstrates strength and support as they are held close by a loving caregiver. With this method they are not alone in their rage, they are safe and kept from hurting themselves or anyone else while they are out of control. Parents need to be physically strong enough to hold the child safely and be emotionally ready to stay calm and loving in the face of intense rage. It is crucial that parents do not get angry while the child is throwing a tantrum. The child must be able to look up into loving eyes and feel safe enough to trust. If parents are not physically strong enough or are not in an emotional position to be able to hold the child without getting angry, parents may choose to leave them lying on the floor alone to scream by calmly walking away or sending the child to their room. When they finish the fit in their room and there is 10 minutes of quiet they then may come out and be held and rocked. At that point processing feelings with them is appropriate with the questions: "What happened? How were you feeling? How did you handle it? How can you handle that better in the future?" These questions are a clear way to process occurrences with them. Feelings must be validated and met with acceptance and love. "I bet you do feel that way." For children who throw many unscheduled fits another option is to require them to have a "practice fit" daily scheduled by a loving parent. If the child is throwing intensely violent fits for longer than an hour at a time, it is recommended that the child be evaluated by a skilled mental health professional.

Respect for Animals

In 1209 A.D. Saint Francis of Assisi wrote, "If you have men who will exclude any of God's creatures from the shelter of pity and compassion; you will have men who will deal likewise with their fellow man." With no conscience development there is little or no empathy for other living things. Supervisory Special Agent Alan Brantley, a psychologist with the FBI in the Behavioral Science unit, has interviewed and profiled numerous violent criminals. In an interview he stated,

> Something we believe is prominently displayed in the histories of people who are habitually violent is animal abuse. Sometimes this violence against animals is symbolic. We have had cases where individuals had an early history of taking stuffed animals or even pictures of animals and carving them up. That is a risk indicator. You can look at cruelty to animals and cruelty to humans as a continuum. Violence against them [animals] indicates violence that may well escalate into violence against humans (Lockwood & Church, 1996, p. 27).

Pets become items to conquer for the child seeking power and control. One little girl being interviewed months after beginning attachment work was asked about the worst things she had done. Her reply was "killing animals." When asked how she felt when killing them, her chilling reply was "happy." Often these young killers do feel joy at the power they feel in the ability to take a life:

> Violent offenders often begin their criminal careers by maiming or killing animals. An FBI study of serial murderers found that most had killed or tortured animals as children or adolescents. All of the alleged young perpetrators in a series of recent schoolyard shootings—in Arkansas, Oregon, Mississippi—were notorious for abusing animals. Kip Kinkel, 15, accused of killing one classmate and injuring 23 others when he opened fire in a high school cafeteria in Springfield, Ore., earlier this year, was known to cut the heads off cats and mount them on the end of sticks. (Warren, 1998, p. 5).

Attachment-disordered children must be closely monitored when in the presence of pets. Animal abuse often starts with teasing or tormenting. If a child is displaying this behavior they must not be permitted to care for, feed, or even interact with animals. "The criminal 'Hall of shame' is filled with people who as children did nasty things to pets. If their parents or teachers had seen the warning signs or known how to counsel them, history might have been different" (Capuzzo, 1994, p. 6). These disturbed children should never be given a pet to "bond with" when they have not bonded to a human. The damage that is done to the child by being allowed to cross that line of abuse makes it much more difficult for them to attach.

In the case of an Attachment-Disordered youngster, petting, controlling, or being the caretaker of living beings must be an earned privilege. It is recommended that a child be well bonded and respectful, responsible, and fun to be around for at least 6 months to a year (depending on the severity of their disorder) before being allowed even well-supervised access to pets. When the privilege has been earned the Mom should hold the child on her lap and support the child's hand as the pet is stroked gently. Guide the feelings by asking questions such as "How do you feel when you are close to this (pup/kitten)?" "How does the pup feel when you do that?" "What are his eyes saying?" "How does his fur feel on your hand?" "Does he like this?" "Does he like you for doing it?" When asked how the child feels being close to a pup or kitten, the often honest reply is "mad" or "like killing." They frequently tap into the unconscious feeling of their own infant rage when they were in a helpless and hopeless state as an infant. These feelings should be validated and met with love and empathy for the child as they continue to stroke the pet. As long as they are releasing these feelings verbally they seldom act on them. Let them talk them out. Parents should listen with empathy and not aghast horror.

Respect for Property

The lines of respect need to include property. Treating furniture and other things in the home respectfully is where a child learns how to treat things in the world. This can deter property damage and vandalism. Children should be expected to sit on chairs with their feet on the floor and without tipping the chair. A defiant challenge on this rule can be handled by having the rule-breaker lose their furniture privilege for the day. An appropriate response is to have the child stand during the rest of the meal if it occurs during a meal or sit on the floor for the rest of the visit or reading time. The consequence should be imposed the first time the rule is broken. Warnings, reminders, and second chances all eliminate the expected consequence. These ineffective actions become a drain on the parents' energy.

In the 1940s some of the top classroom behavior concerns, as reported by teachers, were talking out of turn, chewing gum, walking on the grass, and short skirts. These are all problems of disrespect. As the limits were moved to allow chewing gum, eating during class, no dress code, talking more informally (without hand-raising), and walking on the grass at will, the problems changed. In the 1980s some of the top concerns of teachers involved drug abuse, alcohol abuse, teen pregnancy, and suicide. The limits need to be moved back so that children are testing the mundane limits such as "walking on the grass" rather than testing the "drug laws." If we hold the line on the little things in a loving way, we don't have to deal with the larger ones as often.

Restitution of damaged items is essential for the destructive child to learn to avoid this behavior and feel good about him- or herself. Broken or damaged items that belong to other people must be repaired and/or paid for promptly by hard-earned money on the destroyers part. Over 2000 years ago a wise physician wrote, "Produce fruits to match repentance" (Bible, Luke, 1978). Drawing from birthday or allowance funds seldom has the same impact. Minimum wage should be paid for work done in a timely manner up to the parent's standards. If the assigned task is done poorly or slowly the child may be fired just like in the real world. An unemployed adult in debt does not get the goodies (vacations and cool toys) that a hard-working debt-free person has. A child can learn this lesson early.

A damage deposit is a very powerful tool to help a child who has expensive destructive hobbies. The child must earn a set amount of money to be kept in a fund. The amount to be earned is decided according to the most expensive item the child has stolen or destroyed times two. The total amount must be earned before the child may have any extra privileges. They remain at the lego level until such time. As the child heals and the funds are required less often, a rebate can be set up. Having no need to repay any thefts or damages for a period of 2 months or more, the child may begin to draw a reasonable amount per week.

Responsible for Their Body

Responsibility is an essential skill to survival. Disturbed children often have not been "imprinted" with the patterns that our society has. Patterns such as using the toilet, wiping, flushing, then washing hands; washing your face, brushing your teeth, then combing your hair; taking a shower, washing your hair, washing your body, drying with a towel, then getting dressed. Infants learn these patterns even before they use them by observing adults repeatedly. Disturbed children often do not flush, wash, brush teeth, or dry themselves before dressing. In children over 2 these behaviors must be taught rather than imprinted. It now must become a habit. A morning routine that helps with younger children (under 11) is for Mom to get dressed and use the bathroom, then turn the alarm off the child's door and open it. At that point the child should be up and dressed with their bed made and room picked up. They do not come out of their room in the morning until those three things are done. Children over 5 should have their own alarm clock to get themselves up on time. The reward for being dressed, their bed made and room straightened, is to come out of their room and be with the family. That is the positive reward for effort. They then go in and use the toilet. The Mom and child then prepare for the day together. They wash their hands, brush teeth, wash faces, and comb hair, together. The shared morning routine should be done in the same order together for at least a month, usually 2 months (depending on the child's age), to help them set those patterns.

The Attachment-Disordered child will sometimes stink to get the negative attention they seek and are sometimes filthy because they believe they are trash and it does not matter (they don't care) and/or they want to push the parents' button. Either way it is not healthy for the child emotionally or physically. Some parents choose to have the child don a swimsuit and get in the tub to be scrubbed, "till they shine," by a loving parent. This may continue until the child matures enough to ask to do it on their own. This method has even worked for younger teens. Some odorous youths are confined to the house and yard to save the community from the fumes. The message of unconditional love must be clear. Many people love their horse, dog, or cat no matter what they smell like. A child should be loved even more so. The consequence of paying for drilling and filling decayed teeth has proven successful with some that are toothbrush resistant. No privileges may be earned until the child takes over the tasks of washing, tooth care, and shampooing in an age-appropriate manner.

Many of these children find disgusting hobbies to repulse parents. The hobbies of booger- or scab-collecting, masturbating in front of the family, chewing fingernails, or popping knuckles are all dealt with by prescribing the problem. Setting aside daily time for the child to do these activities in their bathroom or bedroom will often eliminate them. If the parents forget to send the child in for their hobby time, they may request that the child do the particular activity as a reminder for the forgetful parents.

Responsible for Their Possessions

The daily task of bedmaking and placing soiled laundry in it's place is a great start of compliance for the day and a simple demonstration of responsibility for the child. A 2-year-old can assist a parent in this morning task. An older child, when able, should make the bed and put laundry and toys away before being allowed out. It becomes the "passkey" for joining the family for the day in a positive way. If the defiant child refuses to make the bed and pick up laundry, the loving parent makes regular visits, approximately every 30 minutes, to the child's room to hug and let them know they are loved whether in their room or out. It is helpful to let the child know there is no rush. They can come out whenever they have the "passkey" ready. Whatever day that is. The bottom line is that if they cannot make themselves comply with these minor requests the rest of the day will be bad news at school or at home. It is better to be keep them in a safe spot until they are stronger. If the parent must leave, the child may be entrusted to a trained therapeutic respite provider's home to practice making beds for the day. When the gauntlet of defiance is thrown down by the disturbed child the parents must meet that challenge and win it in order for the child to feel safe enough to trust and develop a conscience.

"Attachment-disordered foster children, unable to express anger directly and verbally, often find an outlet in urine and fecal matter" (Delaney, 1991, p. 91). Until they are strong enough to talk it out they will relieve themselves in more odorous ways. This needs to be handled lovingly by parents, allowing the child to clean all their soiled items personally. White vinegar and cold water in a bucket outside works well for this. If the child wants to play hide and seek with soiled clothing, hire another child to sniff out the item. Give the finder a bonus for this chore. These small matters are all tests by the disturbed child to see if the parents possess the strength necessary for the child to heal. If the parents pass the test the child can then move forward in their emotional progress.

Dr. Foster Cline has an excellent illustration of the stages of conscience development in his book *Understanding and Treating the Severely Disturbed Child.*

> All lasting cultures are built upon adequate, pervasive internalization of the Object.... Freud uses the word 'Object' to signify a special internalized person, usually the mother. This person becomes a part of ourselves, and a unity of world and self perception is thus brought about.... Thus, a child goes through predictable stages as he develops his own conscience. In the case of stealing, it might be illustrated as follows:

Stage 1 "I'll take it" (1–3 years old)	Represents primary process thinking. (no lid on the id)
Stage 2 "I would take it, but my dad (mom) would kill me!" Parents are seen as all-powerful in size, power and mental ability ... regardless of how gentle they may be. (2–5 years)	Represents causative thinking although primitive and inaccurate. (Even this type of thinking is not present in children who have severe attachment problems—they steal unless the parent is in sight.) (outside lid on id)
Stage 3 "I would take it but my parents might find out." (still fearful, 5–7 years)	Shows planful causative thinking. The child is "playing the odds." (partial lid on id)
Stage 4 "I would take it, but if my mom found out, she'd be upset." (wistful, 6–9 years)	Shows solid beginning of internalization of object. The child cares about how he perceives his outside "guiding light" as feeling. (Lid on id)
Stage 5 "I would take it, but I don't feel good about doing things like that." (7–11 years)	Internalization is complete and the child's own moral values are "in place." (Superego complete)

(Cline, 1979, p. 88).

Children who do not have parents that they can "perceive as all powerful in size, power, mental and physical ability" commonly do not progress through Stage 2 of conscience development. When they can con and manipulate parents, parents are seen as weaker than the child and not to be trusted and conscience development is derailed. Parents who expect respect and responsibility and impose or allow consequences to occur the first time there is a "test" by the child can make a clear impact on the all important growth of the conscience.

Responsible for Their Job

Kahlil Gibran said, "Work is love made visible." These children need love made very visible: "... good parents give their children appropriate chores to do, and expect the chores to be done well. They do not accept excuses for poor or incomplete work, as this gets the children comfortable with meeting lower expectations" (Cline, 1995, p. 167). If the child is over 2 years old the "job" would be family chores. If they are school-age the job then includes schoolwork and family chores. Helping the family teaches work skills for the real world. It is crucial to their emotional healing that they work to help the family. In 25 years of work with difficult children, I have not seen one child heal without being successful with regular chores.

Doing chores teaches children how to give back. Children can be taught how to "take" very well by constantly being giving to. They must be taught how to give back in order for them to be able to have a loving relationship. General Colin Powell (1997, p. 135), Chairman of America's Promise–The Alliance for Youth, wrote, "Young people—like adults—usually find that when they make a real effort on behalf of others, they get back more than they contribute." Giving back to their family and eventually their community makes a child feel important as a valued member.

Teaching them how to work and work well (to go the extra mile) is how parents can effect student's schoolwork. When they get an "A+" on their household chores, they know how to put out the extra effort to get an "A+" on their science fair project or their schoolwork. It can fine-tune their work ethic. Doing chores teaches life skills. In adulthood the skills to maintain a home and job are essential. When these skills are learned from loving parents, it can be a very bonding and a powerful self-esteem-building tool.

Some appropriate chores for a 2- to 3-year-old would be to sweep porches, fold washcloths, fetch items, match socks, help set the table, fold napkins. More advanced skills such as vacuuming, dusting, floor-scrubbing, laundry-folding, raking, mowing, weeding, and bathroom-cleaning should be added as the child matures. The highest level of chores in most homes is usually dish-washing. Work that should not be included as family chores would be caring for their own body and possessions such as their laundry and bedroom. Each child should be expected to do 30 minutes worth of family chores each day. Younger children can have several small chores throughout the day that equal 30 minutes. The job can be done for a length of time the parents feel is appropriate. The disturbed child can dawdle for hours if given the opportunity to waste time. Productive energy output is the aim. Building the child is the goal. Parents do well to remember the job is not as important as the child.

Parental instruction and supervision of chores must be carried out with high age-appropriate expectations, clear goals, and in a loving way. When they are scrubbing the floor, for example, each corner must be checked. Under the rug should be inspected in case they attempt to cheat. Parents should let them know they care enough about the work that they have done to spend time to carefully look it over and give a big hug for a job well done. Doing a good job (to the level the parent would have done it) teaches children pride in a job well done and builds self-esteem. Quality of effort on chores is a good barometer for the need for more or less frequent therapy sessions. A compliant, eager-to-please child is bonding and growing emotionally and needs less frequent therapeutic intervention than one who is still struggling with defiance and trust.

When a child is being defiant about helping the family with chores it is imperative that no new privileges be given until the child rises above the difficulty.

> Use "thinking" rather than "fighting" words. Make positive statements ("Feel free to join us for lunch when your chores are done"), in contrast to negative statements ("you can't eat until you do your chores"). Thinking words promote positivety and cooperation. Fighting words promote animosity, defiance, and conflict (T. M. Levy & Orlans, 1998, p. 203).

Schoolwork must be solely the child's responsibility. Parents who remind, cajole, and push to get the homework done usually end up with a losing battle on their hands. Pounding knowledge into minds does not work. They must seek it in order for it to penetrate. The school system has built in consequences for homework negligence. Let the consequences fall.

If the child is too disturbed to attend school they should be maintained in a therapeutic environment and a qualified tutor hired to offer opportunities for learning, if the child is responsive. Homeschooling with the mother as a teacher is usually an enormous battle that damages the parent–child relationship further and pushes the child to further reject education or the mother. This is not due to the skill or lack thereof on the mother's part to teach, but rather is due to the pathological thinking of the child to reject the mom and everything she believes.

Emotionally disturbed children have no business being employed outside the home unless the age of emancipation is within 1 year. Then, in preparation for independent living, it is recommended. If college is a clear option, their job, their only job outside the home, should be school until they graduate from high school. Teen jobs often sabotage school work. A study done over a 25-year span on 1700 participants by Delbert Elliot indicates even more problems.

> Teens who work before graduating from high school are 1-1/2 times more likely to commit serious criminal offenses and use alcohol, and are more than twice as likely to experiment with marijuana than those who do not have jobs. . . . If students work grades suffer. Work also reduces the bond to family and school (Elliot, 1992; cited in Brown, 1992, p. 2g).

It is too much cash, too much power, too much freedom, too soon. Small part-time jobs, less than 10 hours a week, *occasionally* still leave time for study and relaxation as well as the time to help the family and participate in church and family fun. Emotionally disturbed children are seldom a benefit to the employer. Being employed outside the home distracts from the healing that must be a priority.

Fun To Be With—Attitude

An attitude of "poor, pitiful me" is often seen in those children attempting to control adults with tear-jerking manipulations. When unsuspecting adults are ensnared by this ploy and mirror the child's eyes full of pity, the message received by the child is that adults are stupid enough to fall for their con; that adults believe that they are poor, poor children; and that adults are against the parents. These are not poor, poor children. They now have a home and loving parents. Parents are obviously caring and committed when they are looking for resources or they wouldn't be reading information and seeking the counsel of wise therapists. They wouldn't be looking for answers. So, to look into that child's eyes with a "poor you" message is not appropriate. Adults need to look into their eyes and express, "Child, you are so blessed to have a mom and a dad like this." Mom and dad's eyes should be greeted with hope and encouragement with the message, "Hang in there. I believe in you." That message of belief in the parents must come through with everything professionals say and do. Maybe the child was abused or left in the past. They are no longer an abused/abandoned child. That is the past. This is the present. To heal, the children need to learn to trust adults. The con must stop.

Triangulation is a common "sport" among children with attachment disorder. They behave defiantly to one adult and victimized to another in order to split the team. Unsuspecting adults begin to mistrust other adults in the child's life. This destructive game can be an attempt to control by dividing and conquering. Mom and Dad begin to doubt each other. Even strangers can be pulled into this game. Be proactive in dealing with this problem. Parents should discuss this with each other, the teacher, the social worker, church members, and relatives. The healthy adults need to believe and trust each other over the tales told by the pathological child. The child should be kept from interacting with strangers and weaker members of the team who may fall for the con game until he or she is stable enough to stop the game. Keeping them home until they earn the privilege to go out by succeeding with all the steps up the ladder for a period of time, such as a month, has proven to be helpful. Having the child wear sunglasses in public has helped to cover the mournful expression they often use.

An attitude of consideration and thoughtfulness goes a long way in establishing and maintaining relationships in life. "Manners" is another word we use for consideration of others, manners such as those used at the table. Eating meals together around the table is an important family time. Family time with *no* television; time together eating, sharing, laughing, and enjoying the food and each others' company. The president of the National Center on Addiction and Substance Abuse at Columbia University, Joseph A. Califano, Jr., stated, "The odds that 12- to 17-year-olds will smoke, drink, or use marijuana rise as the number of meals they have with their parents declines"(1998, p. 4). "Teens who eat dinner with their families regularly fare better than those left on their own," according to a study by *Who's Who Among American High School Students.* Those teens said they felt a direct correlation between the family dinner and feelings of satisfaction and personal well-being. Those who didn't share the family meal were four times as likely to have had sexual intercourse than the group who did eat with their families, the study found. "Students who interact with their family at dinner were twice as likely to spend their time studying" (De Tulrenne, 1996, p. 7D). A child with Attachment Disorder often feels uncomfortable in a close loving setting such as mealtime and will attempt to sabotage it. Eating "gross" at the table or talking with their mouth full is a common occurrence. An effective consequence would be to deal with it the first time they chew with their mouth open. The first time they say one word with food in their mouth, excuse them calmly from the table. Just state, "You are excused." They then may take their plate into the laundry room or out on the picnic table and eat alone—no explanation. If the child is over 2 years old they know exactly what they are doing wrong. They learn from action not words. Action, not anger.

Some children use helplessness as a weapon to control adults. Be aware of those who cannot turn door handles, fasten seatbelts, or use scissors on schoolwork but can pick locks.

Fun To Be With—Gratitude

One of the most important ingredients of a child who is fun to be around is an attitude of gratitude. Children need to appreciate the things that are done for them so they are not wallowing in the "have nots" but are celebrating the "haves." Looking for what they are thankful for is a positive attitude rather than a negative attitude. Parents should model that for them by appreciating each of the things that they do as well as the things parents do for each other. Parents should expect thankfulness. A tool to help with this is the "feelings book." This is a daily journal in which the child writes three feelings they had that day and includes three things they are thankful for. Younger children draw pictures of "the best thing that happened today" and "the worst" as well as what they are thankful for. To make the younger child's "feelings book," five pages of typing/copy paper are folded in half and stapled. This book should be completed daily before the child has dinner. Prepare the meal, serve it, and lovingly cover or refrigerate until the child is ready.

POWER THE CLIMB TO SUCCESS WITH NURTURING

To power the climb, parents must create and maintain with the child a heart-to-heart lifeline using the keys to bonding. These crucial keys are loving interactions required for the development of the mother–child attachment. They include eye contact, skin-to-skin touch, rocking, lactose (milk sugar), smiles, and vocalizations (Ainsworth, 1972; Cline, 1979; Cohen, 1974; Masters & Wellman, 1974; Yarrow, 1961, 1964, 1972). The tight structure and high expectations must be balanced with powerful nurturing. The normal nurturing that occurs during the first year "soul cycle" must be repeated many times during the bonding of an older child as well.

Loving eye contact is a vital, powerful healing connection well.

> Eye contact must be soft and loving. Every minute you spend looking lovingly into your child's eyes will save you about an hour of pain when they're teenagers. "The eyes are the window to the soul." While making eye contact you are giving love from your heart—through your eyes—into your child's eyes—straight into their heart. They will try to avoid it. Be compassionate and firm. They must accept your loving eye contact to heal their heart (N. Thomas, 1997, p. 40).

Seeing is a form of touching at a distance, but touching provides the verification and confirmation of reality. That is the reason why eye *contact* is the perfect example of touching at a distance. Dr. Abraham Levitsky has pointed out that by its very nature, "touch is close and sight is far." We permit contact with those things and people we trust and enjoy. We withdraw from contact with what we don't trust and what we fear (Montagu, 1986, p. 124).

Touch is essential to life. "The communications we transmit through touch constitute the most powerful means of establishing human relationships, the foundation of experience. Where touching begins there love and humanity also begin" (Montagu, 1986, p. xv). "When the need for touch remains unsatisfied, abnormal behavior will result" (Montagu, 1986, p. 46). The angry younger child will often get the touch they need by grabbing another child or hitting or punching. Children need 12 hugs a day during the healing process. They should be hugged when they do a great job to celebrate it. They should be hugged when they do a bad job to cheer them up. Hug them when they are mad, sad, glad, or scared. Hug them just because they *are.*

Teenagers need a tremendous amount of touch. They do fit on a lap in the rocking chair, they just have long legs that hang over the side. Teens are often looking for touch in all the wrong places. Teen sex and pregnancy is rampant in our country. "Many women—especially single ones—become promiscuous to get the holding they want" (Verney & Kelly, 1981) In a study (Hollander, 1961, 1973; cited in Verney & Kelly, 1981, p. 123) on women and holding, over half admitted to having sex to entice a man to hold them. Investigators in a different study (Malinquist, 1966; cited in Verney & Kelly, 1981, p. 123) found that in women who had three or more illegitimate pregnancies almost half revealed that sex was the price they willingly paid to be held. Parents need to hug their teens: 7 hugs a day for maintenance; 12 hugs a day for healing.

Movement such as rocking has been found to be very healing. Rocking motion was found by Dr. Sung Choi, of the Medical College of Virginia, to significantly reduce the time spent in intensive care units. He reports in the *Rocky Mountain News* (4/7/1992, p. 8G) that movement helps reduce the risk of infection and complications from pneumonia and other problems. Rocking is good for mother and child.

> Rocking, in both babies and adults, increases cardiac output and is helpful to the circulation; it promotes respiration and discourages lung congestion; it stimulates muscle tone; and not least important, it maintains the feeling of relatedness (Montagu, 1986, p. 158).

Rocking chairs are comfortable and relaxing for both mother and child. "As the mother gently rocks, she improves the circulation in her legs. The to-and-fro motion stimulates the vestibule apparatus in the child's inner ears, contributing to his better control of balance and position" (Montagu, 1986, p. 166). "I believe that the deprivation of body touch, contact, and movement are the basic causes of a number of emotional disturbances which include depressive and autistic behaviors, hyperactivity, sexual aberration, drug abuse, violence, and aggression" (Prescott, 1985; cited in Montagu, 1986, p. 226).

Smiles are a critical part of connecting. "Smiling and laughter, as Lorenz tells us, are among the tribal signs that unite the members of the human fraternity" (Fraiberg, 1977, p. 47). It is the smile in the mother's eyes that the child connects with. As René Spitz (1965) demonstrated in his studies, the smile is elicited by the configuration of the upper half of the human face.

Sugar is closely interwoven with love in our culture. We call each other endearing terms such as "sugar, honey, muffin, and sweetie" to express fondness for one another. "Among the mammals, human milk is the sweetest of all, containing seven percent milk sugar, compared with four percent in cow's milk" (Montagu, 1986, p. 95). Sugar is an important part of bonding. According to researchers led by Elliott Blass, a psychologist at Johns Hopkins University in Baltimore, infants given sugar before the painful procedures of circumcision and drawing blood by lancing the heal cried significantly less than those given a placebo (cited in *Rocky Mountain News,* 2/6/91, p. 12E). Sugar has been proven, in studies at Duke University by

Richard Surwit and at Vanderbuilt University by Dr. Mark Wolraich, to *not* cause hyperactivity or affect the ability to concentrate. They found "no clear signs that sugar makes children hyperactive" (*Associated Press,* 1994, p. 24C).

A special time of bonding between the child and mother is called "snuggle time." Held as an infant would be, cradled with their eyes locked in to mom's smiling eyes, and with their head in the crook of her arm, they are rocked. During the rocking they can be sung to and/or told silly stories or jokes. Laughter is shared as mom and child enjoy the closeness with caramels, ice cream, or another favorite goody. Snuggle time is a time of sharing the heart-to-heart connection, not a time of lectures. The child should not become a "trapped audience" for the parent to unload onto. It is highly recommended that snuggle time be done 6 days a week, at various times, for about 30 minutes per day. This should continue for 6 to 9 months until the child is filling up with mom's love and then should need to be rocked less often.

We have to have touch to survive. The strong parent in control realizes this and plans on a daily basis to put their arms around the child to hug them, to pat their hands, to look into their eyes lovingly and to spend the time to snuggle them. Loving eye contact, touch, movement, smiles, and sugar all need to be interwoven and connected as part of the heart-to-heart connection, part of reconnecting the ties that bind. The child will try very hard to destroy this essential time together. Powerful parents are prepared and insist on the nurturing.

The disturbed child does not get to select or reject the hugs and snuggle time. "Can I have a hug" does not mean "Mom, I need your closeness and love." It actually translates to "Can I con or coerce you?" If that requested hug is given it implies to the child that adults are too weak to be trusted. The parents must select the appropriate times and ensure the hug and snuggle time quota is completely filled daily.

A Mom that is giving that many hugs and doing snuggle time can become drained. It is recommended to have the Dad recharge the Mom. The recharge is done minute for minute for the time she holds the child. Thirty minutes of snuggle time for the child equals 30 minutes of snuggle time for Mom by Dad. The best thing a man can do for his children is love their mother.

MOVING ONTO STEADY GROUND

As the child reaches for each of the rungs of the ladder and continues the ascent they become stronger. As the child attains new strength he or she becomes more able to handle more privileges. When parents are clear that this child is now on steady ground and ready to move forward with his or her life, then returning some of those items or activities that were enjoyed together is a possibility. Adding them too soon may sabotage their progress.

Attachment-Disordered children are not ready to handle choices until they pass the choice test. They are not told they are being tested. The choice test is to give them a choice between a couple of things such as a drink of milk or water. They will happily choose one of the two or they will choose a third item to let you know they are not ready for choices. If a third item is chosen they are still more interested in a power struggle. At that point, they are not given choices for several months. Until the child becomes stronger the parents should make all the choices.

When the child expresses interest in doing other activities besides chores, legos, reading, art, and mini-trampoline (the only activities they have been doing up to this point), have them make a list of everything they can think of that they may want to do. It should include everything that they can think of on their own, from gum chewing to traveling. Younger children can tell parents and have them write them down. When the list is obtained the parents decide what skills each activity requires. An example would be bike-riding—that would require the child to remain in set boundaries, to know and obey the laws of the road, and have a strong sense of balance (physical/mental development). The child can practice playing legos inside set boundaries for a certain period (determined by the parents) that they feel is long enough to demonstrate the ability to stay within boundaries. They should also display the ability to be responsible by doing chores "fast and snappy and right the first time" rather than "slow and sloppy" for a set period of time. Small activities require less time and more difficult activities require more. For instance, a driver's license should be attained by 1 year of respectful, responsible, fun-to-be-around behavior on a daily basis. Graduation from high school might even be a prerequisite for driving. If they miss one day the time starts over. Set up some goals that only require 1 day, then 1 week, then 2 and so on. A chart should be created and posted for the child to refer to. Owning any kind of weapon should require graduation from college for an Attachment-Disordered child (including a pocket knife). This is a great way to teach goal-setting skills and build the child's self-esteem by cheering them on as they reach each new level of freedom and skills.

Special challenges should be added to the child's life after he or she has moved onto steady ground and is becoming stable. A therapeutic horseback-riding program for emotionally disturbed children has been used very successfully. Wilderness experiences, independent mountain climbing, earning a pet, or whitewater-rafting are all exciting and have successfully lifted a child to greater heights. Another project that has proven very helpful is the Family Virtues Project as described by Linda Kavelin's paper in her *Family Virtues Guide* (Kavelin Popov, 1997).

PATTERNS OF HEALING

As the child is climbing out of the pit they have dug, there are some time bombs that may sabotage them. One of those time bombs is self-esteem. A crucial part of healing, the ability to believe in themselves, grows more slowly than their accomplishments. In the beginning they often have extremely low self-esteem; they believe they are bad, evil, and/or useless. Then they start doing things well, doing their chores well, making good eye contact, speaking respectfully, and getting appropriate positive feedback. They begin to heal emotionally and make faster progress but their self-esteem grows at a much slower rate. Often a child will halt their forward progress and either regress or plateau until their self-esteem growth catches up. This often worries parents. They often begin thinking the plan is not working or that the child has stopped healing or may again get worse. This is a time for parents to catch their breath. They should continue to support the child's self-esteem as they step back to again build up the momentum to move forward. One of the best things parents can do during the healing process, but especially during these lapses or regressions in behavior, is pray. The powerfully effective healing of prayer has been well researched and documented in Dr. Larry Dossey's works (1992, 1993). The studies are fascinating and, more importantly, they provide scientific data to support the hours many parents and grandparents have spent, heads bowed in prayer for these children. It does work. Hearing parents pray, being thankful, especially at mealtime and at bedtime, is an excellent example for children.

> Parental involvement and religious activities are the two most effective protective factors for teens.... Teenagers who attend religious services regularly are far less likely to use drugs, know drug dealers, or have friends who smoke, drink or do drugs than those who attend such services less than once a month (Califano, 1998, p. 4).

Eleanor Roosevelt rightly said, "The future belongs to those who believe in the beauty of their dreams." Self-esteem, defined as who we see ourselves to be, cannot be built until the bond is well established. Words do not change the distorted damaged image inside. These children must see and feel themselves as different before the words mean anything. Self-esteem is the foundation on which a person's personality is built. "There is a direct correlation between low self-esteem and crime, violence, substance abuse, the high school drop out rate, teen pregnancy and other social problems" (*Rocky Mountain News,* 1991, p. 21C). Building it is crucial to success. They then can believe and dream because you do.

> Self-esteem is developed by internalizing the positive feelings projected by a loved one. Acceptance, not just in words but through the eyes, builds self-esteem—rejection destroys it.... Eyes speak volumes to a child. Be aware of what they say! (N. Thomas, 1997, p. 83).

REWARDS AND BONUSES are little surprises given on occasion for above average accomplishments. They are given with love and pizzazz. In the beginning, rewards and bonuses should be food, clothing or necessities. As the child heals and develops better self-esteem, higher levels of bonuses may be awarded. Children with low self-esteem cannot handle a lot of new school clothes or a haul of toys at Christmas and birthdays. To demonstrate this they will quickly destroy them. Keep it simple and have them earn the privilege. Catch the child doing something right! Behavior receiving the most pizzazz is the behavior he will repeat! (N. Thomas, 1997, p. 84).

Avoid labeling (good or bad) "You are lazy," "You are clumsy," "You are wonderful," "You are a terrific kid!" They know they aren't "wonderful" or "terrific" and it makes an adult look like a liar or a fool. They will often blow out for days after one of these positive labels is bestowed on them! They believe they have to prove they aren't "wonderful" or whatever! They usually just agree with the negative labels. Neither one works to build them up to believe in their abilities. DO NOT use them (N. Thomas, 1997, p. 84).

Use positive statements, after the child has shown some effort, with several conditions to build the child's belief in their goodness, ability and appearance. "What a nice smile you have today," "You are getting to be a good helper," "Right now you are doing a good job."

A child who has not internalized a parent stronger than himself can not develop self-esteem. Children with no conscience often have no self-esteem. They will start to take you in when they start to trust you because you are strong (N. Thomas, 1997, p. 84).

They take in self-esteem-building comments much more powerfully if they hear it from parents talking to someone else other than them. An example is, calling grandma to brag about their new accomplishment: "He was quiet for three minutes today!" When they hear parents telling other people positive comments, the child absorbs it more quickly than if you tell them directly.

Anniversary dates and holidays are other time bombs that should be expected. If they've had traumas during their lifetime resulting from moves from one family to another, deaths, surgeries, or abuse that happened at certain times of the year, parents should be prepared for behavioral changes when the anniversary date approaches. Different things may remind them of past pain, such as the first snow reminding them of a trauma that happened during the winter. On an unconscious level, feelings, reflected through behavior, will be stirred up that need to be dealt with by the therapist. Dates of painful events need to be documented as much as possible so all on the team are aware and can be helpful during these times. The child can then overcome that obstacle, climb the wall, and continue forward in his or her progress.

Positron-Emission Tomography scans have shown that we can make physical changes to the brain structure by talking and sharing feelings in therapy. It was reported as "substantial improvement with significant changes in brain function" after only 10 weeks of therapy "this tells us that effective behavioral treatments can have biological effects, not just psychological ones," said Dr. Eric Hollander of Mount Sinai School of Medicine (Hollander, 1996; cited in Goleman 1996, p. 6G). The brain is very elastic and can heal from severe damage as observed in cases of stroke. These children's brains, hearts, and souls can heal from this devastating disorder. It is not an easy undertaking. Dr. Cline believes it takes about 1 month for every year of age of the child. That is time spent creating opportunities for each child to reach for and climb the ladder to success. "It is our firm belief that children hurt by abuse and neglect can learn to love and trust adults in a family setting. Growth and development continue throughout the life span, and it is rarely too late for a child to change" (Keck & Kupecky, 1995, p. 15). "... [Families] invariably share a deep pride and profound joy at playing such an important role in a child's life" (Keck & Kupecky, 1995, p. 179). The children that have ascended from the depths of this disturbance have led the way. Their smiling parents report "*victory.*"

REFERENCES

Ainsworth, M. (1972). Attachment and dependency: A comparison. In J. Ferwirtz (Ed.), *Attachment and dependency.* Cambridge, England: Cambridge University Press.

Associated Press. (1994, February 3). Sugar can't excite kids, study says. *Rocky Mountain News.*

Bible, New International Version. Copyright, 1978 by New York International Bible Society.

Bowlby, J. (1951). *Maternal Care and Mental Health,* World Health Organization. Monograph Series (2), Geneva.

Bowlby, J. (1973). Separation. *Attachment and loss* (Vol. 2). New York: Basic Books.

Bowlby, J. (1977). The making and breaking of affectional bonds: I. Aetology and psychopathology in the light of attachment theory. *British Journal of Psychiatry, 130,* 201–210.

Briggs, B. (1997, April 15). Curtail TV habits, group urges. *Denver Post,* pp. 1E–2E.

Brown, M. (1992, January 2) "Get a job" may not be best advice for teens Surprising study finds kids who work after school more likely to get in trouble with law. *Rocky Mountain News.*

Califano, J., Jr. (1998, November 4). If you eat with your kids they might not do drugs. *The Glenwood Independent,* p. 4.

Campbell, D. (1996). *The Mozart Effect.* New York: Avon Books.

Canfield, J., & Hansen, M. V. (Eds.). (1995). *A Second Helping of Chicken Soup for the Soul.* Deerfield Beach, FL: Health Communications.

Canfield, J., & Hansen, M. V. (1998). *A Fifth Portion of Chicken Soup for the Soul.* Deerfield Beach, FL: Health Communications.

Capuzzo, M. (1994, June 5). Wild things: Animal abuse today, human abuse tomorrow. *Rocky Mountain News.*

Cline, F. W. (1979). *Understanding and treating the severely disturbed child.* Evergreen, CO: Evergreen Consultants in Human Behavior, EC Publications.

Cline, F. W. (1982). *Parent education text: What shall we do with these kids*? Evergreen, CO: Evergreen Consultants in Human Behavior.

Cline, F. W. (1995). *Conscienceless Acts Societal Mayhem.* Golden, CO: The Love and Logic Press.

Cohen, L. (1974). The operational definition of human attachment. *Psychological Bulletin, 81,* 207–217.

Covey, S. R. (1997). *The Seven Habits of Highly Effective Families,* New York: Golden Books.

Crenson, M. (1996, October 14). Kindness toward strangers may be sign of good scents, *Rocky Mountain News,* P. 4A.

Delaney, R. J. (1991). *Fostering changes: Treating attachment-disordered foster children.* Fort Collins, CO: Walter J. Corbett Publishing.

De Tulrenne, V. (1996, January 10). Family sharing at the table isn't extinct, but it has changed. *Rocky Mountain News* p. 7D.

Dobson, J. C. (1987). *Parenting isn't for cowards.* Dallas, TX: World Publishing.

Donovan, D., & McIntyre, D. (1990). *Healing the hurt child.* New York: Norton.

Doskoch, P. (1996, October 9). Humor me: Is laughter really best medicine? *Denver Post,* pp. 1G–2G.

Dossey, L. (1992). *Meaning & medicine: Lessons from a doctor's tales of breakthrough and healing,* New York: Bantam Books.

Dossey, L. (1993). *Healing words: The power of prayer and the practice of medicine,* New York: HarperCollins.

Fraiberg, S. (1977). *Every child's birthright: In defense of mothers.* New York: Basic Books.

Gardner, S. L., Garland, K. R., Merenstein, S. L., & Lubchenco, L. O. (1993). The neonate and the environment: Impact on development. In G. B. Merenstein & S. L. Gardner (Eds.), *Handbook of neonatal intensive care* (3rd ed., pp. 564–608). St. Louis MO: Mosby Year Book.

Goleman, D. (1996, February 15). Brain change tied to therapy. *Denver Post,* p. 80.

Hage, D. (1997). *Therapeutic parenting: it's a matter of attitude*! Silverthorne, CO: Author.

Hollander, M. (1961). Prostitution, the body and human relations. *International Journal of Psychoanalysis, 42,* 404–413.

Hollander, M. (1973). Womens wish to be held: Sexual and non-sexual aspects *Medical Aspects of Sexuality, 3,* 26.

Hughes, D. A. (1997). *Facilitating developmental attachment.* Northvale, NJ: Jason Aronson.

Kavelin Povov, L. (1997). *The Family Virtues Guide,* New York: Plume.

Karen, R. (1994). *Becoming attached.* Oxford & New York: Oxford University Press.

Keck, G., & Kupecky, R. (1995). *Adopting the hurt child.* Colorado Springs, CO: Piñon Press.

Levy, D. (1943). *Maternal overprotection.* New York: Columbia University Press.

Levy, T. M., & Orlans, M. (1998). *Attachment, trauma, and healing.* Washington, DC: CWLA Publications.

Lockwood, R., & Church, A. (1996). Deadly serious. *The Humane Society of the United States News, 41* (4), 27–30.

Malinquist, C. (1966). Personality characteristics of women with repeated illegitimacies: Descriptive aspects. *American Journal of Orthopsychiatry, 36,* 476.

Masters, J., & Wellman, H. (1974). The study of human infant attachment: A procedural critique. *Psychological Bulletin, 81,* 218–237.

McKelvey, C. (1995). *Give them roots, then let them fly: Understanding attachment therapy.* Kearney, NE: Morris Publishing.

Montagu, A. (1986). *Touching: The human significance of the skin* (3rd ed.). New York: Harper & Row.

Pickle, P. (1997a). *Field manual: Basic training for parents.* Evergreen, CO: The Attachment Center at Evergreen.

Pickle, P. (1997b). *Life in the trenches: Survival tactics.* Evergreen, CO: The Attachment Center at Evergreen.

Powell, C. (1997, December 15). Everybody's children. *Time,* p. 135.

Prescott, J. H. (April 1975). Body Pleasure & the Origins of Violence, *The Futurist,* pp. 64–65.

Randolph, E. (1997). *Children who shock and surprise* (Rev. ed.). Kittredge, CO: RFR Publications.

Reuter, (July 29, 1994). Studies link memories with sleep. *Rocky Mountain News.*
Rocky Mountain News. (February 8, 1991). Some don't feel good about self-esteem bills. *Rocky Mountain News.*
Rocky Mountain News (January 25, 1994). Turn off TV, pope tells parents, pp. 54A.
Rocky Mountain News (April 7, 1992). Rocking for Health, pp. 8G.
Rocky Mountain News (February 6, 1991). How sweet it is, pp. 12E.
Shapiro, F. (1995). *Eye movement desensitization and reprocessing.* New York: Guilford Press.
Spitz, R. A. (1965). *The First Year of Life,* New York: International Universities Press.
Thomas, E. (1996, April 22). Blood brothers. *Newsweek,* pp. 28–38.
Thomas, N. (1997). *When love is not enough: A guide to parenting children with RAD.* Glenwood Springs, CO: Author.
Verny, T., & Kelly, J. (1981). *The secret life of the unborn child.* New York: Dell.
Verrier, N. N. (1993). *Primal wound.* Baltimore, MD: Gateway Press.
Warren, J. (1998, August 11). Murderers often started on animals. *The Glenwood Independent,* p. 5.
Williamson, M. (1992). *A return to love.* New York: HarperCollins.
Yarrow, L. (1961). Maternal deprivation: Towards an empirical and conceptual re-evaluation. *Psychological Bulletin, 58,* 459–490.
Yarrow, L. (1964). Separation from parents during early childhood. In M. Hoffman & L. Hoffman (Eds.), *Review of child development research,* New York: Russell Sage Foundation.
Yarrow, L. (1972). Attachment and dependency: A developmental perspective. In J. Gerwitz (Ed.), *Attachment and dependency.* Washington, DC: V. H. Winston.

BIOGRAPHY

Nancy Thomas

Nancy Thomas, the mother of three birth children, one adopted child, one child being raised under legal guardianship, and grandmother of six, has been a Therapeutic Parenting Specialist since 1985. Nancy has worked as cotherapist with children in intensive attachment therapy with many renowned experts in the attachment field. Nancy worked as secondary lay-therapist with Connell Watkins, M.S.W., for over 2000 hours with children that had symptoms of being borderline psychotic, ritualistically abused, abusive to animals, mood disordered, ADD, ADHD, Reactive Attachment Disorder, and having Tourette's Syndrome. Many had histories of homicide, arson, and/or sexual perpetration.

Nancy trained and worked with Foster Cline, M.D., psychiatrist and founder of The Attachment Center at Evergreen, who has praised Nancy as "the foster parent of foster parents" (keynote address, National ATTACh Conference, 1997). Dr. Cline said of Nancy, "She has been acclaimed by national as well as local audiences. She has taught both professionals and front line workers." Nancy spent several years under the mentorship of the top therapeutic mom at the Attachment Center at Evergreen, Lori Wilson.

Nancy and her family live in the high mountains of western Colorado in a home centered around healing the hearts of children. Nancy and her husband, Jerry, have shared their life and home for over 25 years with severely emotionally disturbed children with attachment problems. Ninety percent of the children placed in her care are kids who had killed. She has an 85% success rate with these high-risk children which is one of the highest in the country. She specializes in bonding and conscience development.

Based on her years of hands on experience and high success rate Nancy has authored a parenting guide, entitled *When Love Is Not Enough: A Guide to Parenting Children with Reactive Attachment Disorder,* which has been well received by parents and professionals. Recognized internationally as one of the leading authorities on parenting emotionally disturbed children, Nancy was asked to join several esteemed attachment professionals in coauthoring *The Handbook of Attachment Interventions.*

Nancy was honored in a front-page article by the *Glenwood Post,* April 12, 1997, entitled "*Using Love, Woman Opens Heart To Children Of All Types.*" The publication of the American Psychological Association, the *APA Monitor* in the June, 1997 edition, quoted Nancy along with top specialists of Attachment Disorder in an article entitled "*When Children Don't Bond With Parents.*" The Cline/Fay Institute (the Love and Logic folks) recently published her audio tapes "*Healing Trust: Rebuilding The Broken Bond.*"

Since her work was highlighted in an HBO special in 1990 entitled "Child of Rage" she has become a much-sought-after speaker.

Currently, she spends her "spare time" teaching Power-Parenting classes for parents of children out of control.

4

INTEGRATING ATTACHMENT CONCEPTS FROM WESTERN PSYCHOLOGICAL AND BUDDHIST PERSPECTIVES

BONNIE RABER WICKES

Private Practice
Anchorage, Alaska

Attachment is a topic of clinical interest that has profoundly influenced thinking about personality development and functioning in Western psychology over the past several decades. The concept of attachment has long been a focus of Buddhist teachings as well. The purpose of this chapter is to discuss and integrate Western psychological theories and Buddhist views of attachment as they relate to this author's evolving personal theoretical orientation and psychotherapy practice.

Attachment theory (Bowlby, 1969, 1973, 1980) assumes that the developing infant's early attachment-related experiences are in time represented cognitively in the form of internal working models of self and other, which in turn affect personality functioning and relationships throughout life. Kagan (1996) identified research on infant attachment as the most contemporary form of the popular hypothesis that the experiences of infants can create schemata, habits, and emotions that are enduring. Kagan concedes that the attachment bond is a useful construct. However, he believes the inference of stability over time, of behavioral and affective structures, rests on a more fragile foundation. Buddhist ideas about attachment and self might speak to the impermanence implied by Kagan.

While Western psychological theory focuses on how attachment, separation, and loss affect development, the Buddhist concept of attachment and aversion focus on one's characteristic interaction with experience (Fontana, 1990b). From the Buddhist perspective, aversion is actually negative attachment. According to Fontana, our freedom is limited because we are constantly craving after our attachments and seeking to escape from our aversions, thinking that will make us happy. This keeps us from accepting that external reality, like the internal reality of the self, is ever-changing, or impermanent. In this view, attachment promotes suffering, since the things to which we are attached inevitably change and pass away, leaving us to grieve the loss.

There is a long history (in Epstein, 1995) of at least a potential marriage between Western psychology and Buddhism: William James predicted that Buddhism would be a major influence on Western psychology; Jung was interested in and influenced by Eastern mysticism during his lifetime; the influence of Eastern thought may be seen in the work of Abraham Maslow; and Erich Fromm and Karen Horney were both attracted to Buddhism late in their careers.

Rinzler and Gordon (1980) describe most early attempts, however, to wed psychotherapy and Buddhism as "shotgun weddings" (p. 52), concluding that psychotherapy and Buddhism view human existence so differently that drawing parallels or achieving synthesis is almost impossible. They acknowledge that each began as a response to human suffering, but point out that psychotherapy assumes the individual can be changed so pain is minimized, while Buddhism regards the very notion of individuality as the cause or source of pain. Buddhism does not attempt to change our ego structure, as Western psychotherapy might, but questions the solidity and reality of the structure itself.

Although the psychological clarity of Buddhist teachings has been respected by many of the early pioneers in Western psychology, their relative inexperience with ***both*** clinical psychotherapy **and** intensive Buddhist meditation has hampered the development of an effective integration of the two (Epstein, 1995). Today, there are a growing number of psychologists, psychotherapists, and other mental health professionals who bring Buddhism into their work or work psychologically from a Buddhist perspective (e.g., Epstein, 1995; Joyce, 1997; Kabat-Zinn, 1990; Kornfield, 1993; Linehan, 1993; Rubin, 1996).

THEORETICAL FOUNDATION, ASSUMPTIONS, AND PRACTICE ORIENTATION

Attachment Theory

Attachment theory grew out of early object relations thinking (e.g., Klein, 1952), studies of the effects of maternal deprivation on personality development, and research in ethology. Bowlby (1969, 1973, 1980) developed attachment theory as a way of conceptualizing the natural tendency of human beings to make strong bonds to particular others for the purpose of protection and survival. He addressed the evolutionary and relational quality of personality growth and development. Bowlby refers to object relationships as ***affectional bonds.***

Attachment theory shares commonalities with self-psychology (Kohut, 1977). In a review of object relations theories, Bacal and Newman (1990) suggest that the essence of attachment theory and self-object theory is the same: each focuses on the central importance, for healthy development, of the link with a phase-appropriate responsive and supportive figure. The need for the self-object and the attachment figure is not regarded as pathological by either theory and is viewed as legitimate throughout the life cycle. While Kohut emphasized the particular functions provided by the self-object, Bowlby focused on the phenomenon of attachment to a particular person or figure.

Horowitz (1988) joined object relations theory to cognitive science and theory by describing internalized object relations as interpersonal schema. Constructivist cognitive theorists (Guidano & Liotto, 1983; Mahoney, 1990; Safran & Segal, 1990; Varela, Thompson, & Rosch, 1995) suggest that cognitive developmental processes occur within an interpersonal context and have found attachment theory compatible with cognitive science principles. The regulation of attachment behavior is conceived of in terms of a complex feedback system. Attachment behavior is viewed as a wired-in action pattern of fundamental significance in human beings (Bowlby, 1979; Sroufe, 1979; Stern, 1985).

Separation and loss are experiences inherently addressed by attachment theorists. Unwilling separation and loss give rise to many emotions including anxiety, anger, sadness, depression, and emotional detachment. The fear of separation from the attachment figure, and how separations are handled, play a significant role in the development of the child's emotional life. According to this theory, various psychological problems are the result of disturbed attachment. Securely attached individuals operate out of the love and warmth that is generated in the attachment process. Ainsworth (1973, 1979) speaks of the ***attachment–exploration balance*** and the ***secure base phenomenon,*** whereby the attachment figure in time somehow becomes internalized and serves as a base or foundation from which to explore and operate in the world.

Attachment in Buddhism and Mindfulness Meditation Practice

The Buddhist Wheel of Life offers a picture of human existence that Epstein (1995) calls "a Buddhist model of the neurotic mind" (p. 15). The Wheel encompasses six realms of existence through which sentient beings are believed to cycle endlessly in their round of births, deaths, and rebirths.

Epstein views this wheel as a good starting point in comparing Buddhism with psychotherapy. Each realm becomes a metaphor for a different psychological state and is regarded as a particular cause of suffering that arises when human beings are driven by greed, delusion, and hatred (p. 16).

Attachment is a significant feature in this conceptualization. There are twelve interdependent links or stages in this circular chain of life and experience. ***Attachment*** (Gyatso, 1992) or ***craving*** (in other translations, e.g., as cited in Low) is the eighth link. The circular chain is referred to in Buddhism as the cycle of Dependent Origination. ***Dependent origination*** is the notion that on the basis of this present moment of experience, the next moment of experience will arise. This is not a simple linear process of cause and effect, but a spiral of increasingly complex interactions (Low, 1990).

It is at the stage of attachment/craving that some Buddhist traditions say the Buddha formulated the technique of mindfulness meditation as a way to break the chain of past conditioning (Varela et al., 1995). Mindfulness meditation is the hallmark and foundation of all Buddhist traditions. While many different kinds of attachment or craving have been identified, its basic form is desire for what is pleasurable and aversion for what is displeasurable. Theoretically, attachment/craving, the eighth link, arises from feeling, the seventh link, and gives rise to grasping, the ninth link. The concern of the meditator is to break the wheel of conditioned origination and become aware. Awareness brings wisdom and inner peace.

According to Buddhism, the first seven stages of this circular chain of life are linked together automatically on the basis of past conditioning. At the point of craving or attachment, the **aware** person can do something about the future, either interrupting the chain or letting it go on the next link (Varela et al., 1995). **Awareness** is the result or outcome of being mindful. By precise, disciplined mindfulness to every moment, one can interrupt the chain of automatic conditioning of the past. Doing nothing, nonaction, or simply sitting with a feeling, brings eventual awareness of the insubstantiability of it and is enough to break the chain. Nonaction is more difficult than it sounds. Interruption of habitual patterns results in further mindfulness, eventually allowing the relaxation into deeper awareness. For experienced practioners, this brings even more open possibilities in awareness and the capacity to develop insight into the arising and subsiding of experienced phenomena.

Drawing on the results of over 50 studies, social psychologist Ellen Langer (1989) describes the nature or quality of mindfulness: (a) the creation of new categories, just as mindlessness is rigid reliance on old ones; (b) openness to new information and (c) to different points of view; (d) increased control over context; and (e) a process, rather than outcome, orientation. While Langer draws similarities to various concepts of mindfulness found in Eastern traditions, her research has been conducted almost entirely within the Western scientific perspective. She explores the potential benefits of a mindful attitude in aging, health, creativity, and career.

Western Psychological and Buddhist Views of Self

Notions of ***self*** are central to psychological theory and vary widely within Western psychology and across cultural and world views. Generally, however, Western psychodynamic psychology has viewed the self in terms of a mental structure, often referred to as **ego,** which develops in response to early childhood experience. From an attachment theory perspective, early bonding and attachment-related experience is crucial in the formation of this structure. Psychological concepts of the self appear to be slowly changing from a "structure" to a "pattern": For example, the self, according to Stern (1985), is an invariant pattern of awarenesses through which psychological processes are organized, and ego-self is the historical pattern among moment-to-moment emergent formations (Varela et al., 1995).

The stability of this structure, or pattern (called self), is relevant to the integration of attachment concepts in Western psychology and Buddhism. Rosenbaum and Dyckman (1995) challenge Western philosophical and psychological models of self as containers of experience where traits, objects, and memories are organized into stable internal representations reflected in thought, feeling, and action. This view of self, they argue, puts self into a category such as a ***thing,*** seen as unitary, rather than as a composite experience. The self is not an accrual of experience, they suggest, but an ongoing, ever-changing manifestation of potentiality, a ***process.***

Similarly, the Buddhist view is that a solid unitary self is an illusion, sometimes referred to as ***no-self*** or ***empty self.*** According to Fontana (1990a), the doctrine of ***annata,*** empty self, is designed simply to present the self as a dynamic, fluid process rather than a rigid, static structure. This doctrine sees the self as continually subject to change, modified and reshaped by each moment of experience, thought, and action. It is the ***experience*** of mindfulness meditation that defines or reveals the Buddhist notion of self (Kornfield, 1993). Paradoxically, inherent in the recognition of empty self is the experiential awareness, based on mindfulness practice, that the true nature of moment-to-moment being is openness, connectedness, integrity, and belonging. This is often referred to as Self rather than self, and is already and ever present.

V. Brown, a psychiatrist who has also been a long-time meditation practitioner, suggests that it is this moment-to-moment Being, or essence, experienced in a shared and reciprocal manner, by mother and infant that is perhaps the source of healthy bonding and attachment (personal communication, October 30, 1998). Here, the caregiver is in a meditativelike state, responding to this pure essence, or moment-to-moment Being, that is so available to the infant, who is not yet socialized with language and other aspects of cognitive development.

At the core of constructivist theory is a view of human beings as active agents who, individually and collectively, coconstitute the meaning of their experiential world (Neimeyer, 1993). Constructivist notions of self are congruent with the notion of self as **process.** Maturana and Varela (1987/1992) remind us that we are forever experiencing a world and that we cannot separate our history of actions from how this world appears to us in the present moment. They suggest that in the intimacy of recurrent interactions, which personalize the other individual with a linguistic distinction such as a **name,** the conditions may have been present for the appearance of a **self.** Furthermore, they define **love** as the act of one individual seeing the other person and opening up room or space for existence beside him or her.

This seems to be a positive expression of attachment. It is opening up space beside oneself for the other. The healthiest attachments, throughout the life cycle, are characterized by nonattachment as well—space, exploration, trust, and room to grow. Attachment, as a process at its best, recognizes and honors the natural letting go or nonattachment as part of the complete or whole process. Perhaps, then, the self as process is both full, reflecting attachment, and empty, reflecting nonattachment.

VISION OF THE HUMAN CHANGE PROCESS

Attachment Theory and Change

Attachment theory has been advanced as a sociobiological theory, drawing on the premise that human beings are biologically wired to develop affectional bonds for the purpose of protection and survival (Bowlby, 1979, 1982). Attachment theorists acknowledge that substantial psychological change is possible, for example, when children's environments change (Carlson & Sroufe, 1995; Sroufe, 1996). A recent comparative psychology analysis tells us that despite differences in development, form, and biological functions, primate emotional attachments, whether parental or filial, appear to be linked to the same psychobiological core (Mason, 1997).

Main, Kaplan, and Cassidy (1985) define the internal working model of attachment as a set of conscious and/or unconscious rules for the organization of information. According to Main and her associates, internal working models show a strong propensity for stability, but they are not conceived as templates and can be changed. Reconstruction is difficult since internal working models, once organized, tend to operate outside conscious awareness and resist change. Psychotherapy can be a process for restructuring and/or recreating new internal working models from which to draw. Awareness, or bringing into consciousness those unconscious rules for the organization of information, is vital to the psychotherapy change process.

Resistance to change is common in psychotherapy. Mahoney (1990) offers different explanations for resistance to change, including self-protection theories offered by humanistic, existential, constructivist, and self-psychology approaches. Self-protective theories uniquely emphasize the client's healthy caution about embracing experiences that challenge their integrity, coherence, or viability as a living system. Noticing, honoring, validating, and understanding a client's resistance are important to the therapeutic change process and congruent with both attachment theory and mindfulness practice.

Constructivism, Choice, and Change

Neimeyer (1993) describes constructivism as the philosophical framework that emphasizes the self-organizing and proactive features of human knowing and their implication for human change. The constructivist perspective views problems as episodes of disorder that reflect discrepancies between environmental challenges and the individual's present capacities rather than as dysfunctions or deficits to be controlled or eliminated. For the constructivists, the therapeutic relationship provides a safe, caring, and intense context in which the client can explore and develop relationships with self and the world, thus bringing about change.

The experience of mindfulness practice is that change is always occurring in our thoughts and feelings. Nonattachment, the space between thoughts and between feelings, gives one an opportunity to experience the true nature of Self or Being as empty and spacious, on the one hand, and as deeply connected and interrelated, on the other. Awareness gives us the choice to change. Physicist and award-winning author Gary Zukav (1990) places ***choice*** at the center of the evolutionary process.

Is it necessary that one have a cohesive, stable sense of self in order to experience the true nature of Self, where there is no judgment, no striving, no attachment, craving, or grasping? Fontana (1990a) addresses this question about change in terms of ***self-assertion*** and ***self-transcendence.*** Through self-assertion, often a goal in therapy, one builds up the concept of a separate, real, individual self that seeks to maintain itself in the face of the various levels of challenge presented by the external world. The achievement of self-assertion, he says, is a prerequisite for further psychological and spiritual growth, involving a process of self-transcendence.

Kornfield (1993) addresses this issue by posing two parallel tasks in spiritual life: one is to discover selflessness, the other is to develop a healthy sense of self. Both sides, he says, of this paradox must be filled for us to awaken. While it is natural to think of this as a linear process, it may be a nonlinear journey. In attachment terms, it might be that we need to achieve, or experience, a certain amount or quality of attachment before we can transcend, or transform, to experience nonattachment, **annata,** or empty self.

Motivation to Change

It seems the therapeutic change process is usually motivated by something, an inner yearning or desire to be different, or a resounding cry or hurt that becomes intolerable. Outside forces can motivate change as well. Sometimes individuals do not want to change, but are ordered to attend psychological services in the hopes that we might instill in them a desire for change, often for the sake of others or society at large. Change can also occur without consciousness of its happening, so that the yearning and desire may be very deep, unacknowledged, out of awareness. Bringing this desire into awareness is often a part of the psychotherapy process. Change might happen because someone or something else has changed. This is a concept in modern physics that is not fully understood.

Motivation may have momentum all its own, but, generally, change occurs (or does not) in the context of support or reinforcement at many different levels: ***biologically,*** it may be neuronal and cellular, dependent on genes, organs, chemistry, nutrition, touch; ***psychodynamically,*** it may be developmental, archetypal, transferential, dependent on love, trust, respect, attunement, empathy; ***environmentally,*** it may have to do with the air, food, land, home, space, education; ***cognitively,*** it may have to do with language, meaning, understanding; ***socially,*** it may involve friendship, career, play, culture; and ***transpersonally,*** it may occur in the context of nature, solitude, prayer, meditation, music, metaphysics. Change is a complex phenomenon.

Specific symptomatic change is most often what individuals seek in clinical settings. They may seek relief from depression, anxiety, loneliness, self-destructive or addictive behaviors, sexual problems and difficulties, relationship troubles, physical illness, or thought-disorders. From an attachment perspective, one might ask not only what is the individual's attachment history, but to what beliefs, ideas, behaviors, and objects is this person attached and why? What is reinforcing these attachments? What stands in the way of letting go of unhealthy thoughts, beliefs, behaviors, and relationships?

PSYCHOTHERAPY FROM AN ATTACHMENT PERSPECTIVE

The task of therapy, in attachment terms, involves creating healthier attachments for a more functional flow of emotion, thoughts, and behaviors. Farber, Lippert, and Nevas (1995) evaluate the therapist as an attachment figure in terms of being a secure base for exploration, an insurer of survival, and an object of intense affect. While commonalities in dynamics between parent–child and therapist–client relationships may exist, the therapy relationship is mediated by unique temporal, financial, structural, and ethical boundaries.

The idea of therapist as attachment figure in the lives and minds of clients works at different levels. In its simplest form, this orientation means we make it safe to explore new ways of being and thinking about life and the problems and challenges in living. At a deeper level, we might be viewed and experienced as (self-objects) or as attachment figures with the potential for internalization, creating new mental representations and experiences from which to draw.

With mindfulness practice as part of this intervention, the client (and therapist) are continually reminded that change is not only possible, but inevitable. Psychotherapy provides a fluid path to cocreate solutions to complex human conditions and problems. The therapist not only provides safety and nurturing through an empathic attunement and attachment process, but challenges old belief systems and old patterns of viewing and being in the world.

Implications for Clinical Practice

There are several obvious implications of this theoretical integration of attachment concepts from Western psychology and Buddhism for clinical practice. First, a "both–and" rather than an "either-or" approach to healing must be assumed at different levels. The **either–or** split is a common feature of Western thinking and has been evident in creating such distinctions in psychology as the nature-versus-nurture debate. The "both–and" conceptualization was introduced by de Shazer (1985) and has been used since in the synthesis of various theories and approaches. Awareness of **both** the importance of attachment relationships **and** the ability to detach and see ourselves apart from our attachments is equally important in the healing process. Another level implied by this integration is that psychological change involves **all** aspects of the self, including physical, emotional, cognitive-behavioral, and spiritual or transpersonal.

Second, experiential work is a necessary part of clinical intervention based on this integration. Mindfulness practice is a skill introduced for developing awareness. Meditation is but one way to teach or learn this skill. Other ways of achieving mindfulness include journaling, paying attention, noticing, and decreasing mindless activities. The therapist/teacher/model/facilitator/consultant is possibly more effective if he or she understands mindfulness experientially. Some would say this is essential, just as some believe that having experienced psychotherapy as a client is essential for therapists to do their work effectively. Awareness of one's attachment history is also an important part of clinical intervention from this perspective.

Third, the attachment phenomenon extends beyond attachment to other(s). Attachment to thoughts and ideas, feelings and emotions, behaviors and habits, objects and things are a natural part of living in this world. Separation, loss, detachment, and letting go are also a natural part of living. The concept of attachment is inclusive and broadened in this perspective.

Finally, nonattachment, like attachment, is a process through which one can see more clearly what the self is and is not. Western developmental psychology brings us an understanding of healthy attachment; Buddhist psychology brings us awareness, through noticing, moment to moment, our attachment and nonattachment processes, of the inevitability and possibility for change. A hologram might best represent this integration in a symbolic way.

Included in the Appendix is an outline of one way to introduce attachment concepts in a psychotherapy setting. While this outline was developed by this author primarily for group therapy, it has been used successfully as well in individual therapy as a hand-out for information, discussion, and homework. There are many less structured ways to include this theoretical integration in the therapy context, allowing it to unfold as therapy proceeds and attending to individual differences with a focus on attachments to others, to ideas and behaviors, grief and loss issues, mindfulness, and other avenues toward awareness.

Ethical Considerations

Both personal and professional ethics are guiding principles in clinical practice. Personal ethics have to do with taking care of oneself, doing (and being) one's best, keeping an open heart and mind. Professional ethical principles and standards are important and necessary because of the power and influence therapists have in the lives and minds of individuals who seek therapy. These standards are also important in establishing, maintaining, and furthering psychology as a helping and healing profession. Respect for and sensitivity toward one's cultural history and belief system is a starting point for each therapeutic encounter.

Some individuals come to therapy with preconceived ideas about mediation and it is important to be sensitive to this. The fact that mindfulness meditation models are now being taught for stress reduction and pain management at major medical and mental health centers is helping individuals to accept Buddhist practice in Western healing paradigms. While mindfulness practice came out of Buddhist tradition, Kabat-Zinn (1990) points out that mindfulness is basically just a particular way of paying attention and that it stands on its own as a powerful vehicle for self-understanding and healing.

Individuals seeking therapy should have informed **choice** regarding the different levels and approaches to treatment. Questions and concerns about duration of treatment, success rates, and options and alternatives all need to be addressed. Symptom relief may be achieved through medication alone. Brief or long-term psychotherapy, on its own or in conjunction with medication, can also be a rich transformative healing process, resulting in deep and profound lifestyle and/or personality change. Depth psychotherapies transform the basic dynamics from which we operate in the world. Eliade (1969), a yoga scholar, tells us that life cannot be repaired, it can only be recreated. In a similar vein, depth psychotherapy is a recreation process which comes through awareness, understanding, and practice. Working psychotherapeutically from an integration of attachment theory and Buddhist concepts might be considered a depth psychotherapy. It addresses not only the dynamics and belief system one is born into or reared in, but the belief systems of differing world views as well.

Competency, integrity, and responsibility are guiding principles in psychological work, no matter what orientation is chosen. Issues like confidentiality, forensics, dual relationships, and misrepresentation of services often require familiarity with local statutes in addition to the ethical principles and standards established by the American Psychological Association ("Ethical Principles," 1992). Duty to warn, malpractice, and child abuse reporting are examples of ongoing ethical and legal issues. Consultation and/or supervision are important, not only for beginning therapists, but for experienced therapists as well. Referral to other trained professionals requires knowing the scope of services available.

CRITIQUE AND GENERAL REFLECTIONS

Attachment, separation, and loss may indeed be conditions of living, and it is these experiences that Bowlby focused on in the development of attachment theory. While the integration of Buddhist principles of attachment and nonattachment may expand the interpretive possibilities, it probably is the case that not all psychological problems and issues are rooted in attachment-related experiences. This discussion has not been an attempt to claim that they are, but rather to portray an integration of Western psychodynamic and cognitive theories with Buddhist teachings about attachment and nonattachment.

A limitation of the integration of Western and Buddhist models is the difficulty, in general, of integrating approaches that are based on differing assumptions. Paradoxically, this may also be a strength, depending on one's experience and ability to understand and hold contradictions. The difficulty I have experienced in trying to articulate what seems so clear in my mind and experience of psychotherapy and meditation practice might speak to my own gaps in understanding Buddhism and/or Western psychological theory, as well as this general difficulty in integration of theories arising from separate world views and assumptions.

Attachment theory has been challenged by feminist theorists for the central focus on the mother's role in creating health and psychopathology in her child. Attachment theorists (e.g., Holmes, 1993) argue that infants sometimes attach primarily to the father, or another caregiver, and that it is society, not Bowlby's theory, that assigns the role of attachment figure primarily to the mother. While not the subject of this chapter, it is interesting to ponder gender differences from an attachment perspective. Feminist theorists (in Brown & Ballou, 1992) have concluded that males and females are socialized differently and therefore experience relatedness and separateness in different ways, resulting in females defining their self in relation and males defining their self in separation. Sociocultural factors and context play a part in the attachment process, and Western psychological theory is only now beginning to address these complex factors.

The most striking reflection, and affirmation, upon completing this discussion comes by way of seeing and hearing His Holiness the Dalai Lama of Tibet (Gyatso, 1997) speak to a large audience recently on the subject of cultivating inner peace. He spoke of the importance for children to grow up with self-confidence, in an affectionate atmosphere, giving credit to his mother for instilling in him the values of patience, gentleness, and kindness. I thought to myself, "His Holiness sounds like a Western psychologist!" He was very clear, however, that from a Buddhist perspective, it is possible for everyone, no matter what the early experience has been, to know happiness and inner peace because essential human nature is that of affection, gentleness, kindness, and compassion. This nature, he said repeatedly, is already and ever present, and it is through mindfulness to each and every moment that we come to know this.

In summary, both Western psychological theory and Buddhism offer paths toward wholeness and mental health by focusing on different sides, angles, and dimensions of the attachment experience. Western theory addresses the relationship, particularly in early life, and its profound influence on the developing psyche of the individual. Buddhist notions of self offer an expansive view of change, meaning, and possibility; mindfulness practice leads to awareness, experiencing life as it is in the present moment. The integration of both Western psychology and Buddhism into the psychotherapy process can enrich and deepen our understanding of who we are and heighten our awareness of who we are becoming.

APPENDIX

SELF-ESTEEM, ATTACHMENT, AND NONATTACHMENT

Attachment is a term that is often used in psychology to refer to our early development, so it might be important to talk about it in our study of ***self-esteem.*** Our parent, usually our mother, is commonly referred to as an early ***attachment figure.*** An attachment figure could be any primary care-giver—father, babysitter, grandparent—anyone who spends large amounts of time with the developing child.

It is generally believed that if the **attachment figure** is emotionally healthy, chances are pretty good that the child will grow up with positive, healthy self esteem. We can think of it in this way: the child, if it has opportunities to **attach** herself/himself to healthy figures, will learn to have healthy thoughts and feelings about who he/she is. These thoughts and feelings will become **internalized.**

Internalization is important since self-esteem comes from within. This happens in different ways. For example, watching or observing an attachment figure and in time identifying with him or her is a way to internalize a sense of self. Another way is through messages that are either spoken or unspoken. If the attachment is toward an unhealthy figure, he/she learns unhealthy or negative self-esteem. Realistically, most people are a mix of healthy and unhealthy emotions, thoughts, and behaviors.

EXERCISE 1

Think about people in your early life, the first 5 or 10 years. Who do you think your primary attachment figure might have been? Any other important attachment figures? Make a list, including no more than three important people in your childhood and write healthy and unhealthy behaviors that might have affected your developing self-esteem:

Attachment figure	Healthy behaviors	Unhealthy behaviors
________________	________________	________________
________________	________________	________________
________________	________________	________________

As we grow up, and our minds and bodies develop, we not only **attach** ourselves to people, we attach ourselves to things, ideas, thoughts, behaviors, objects. It is somewhat natural to do this. These **attachments** can be positive (healthy) or negative (unhealthy) or somewhere in-between. Here are some examples:

I'm attached to the idea that I can always change to make things better for myself.
I was attached to my therapist when I was in therapy.
I'm attached to watching "ER" every Thursday night.
I used to be attached to smoking cigarettes, but not anymore.

EXERCISE 2

Let's hear some of your attachments (is this attachment healthy, unhealthy, or somewhere in between?):

Thoughts/beliefs +/− Behaviors +/− Objects +/−

Nonattachment is a state of being unattached and is just as important in understanding self-esteem as **attachment. Detaching, letting go, turning it over, gaining some perspective, distancing oneself, needing and creating space** are all expressions that describe a process that leads to **nonattachment.**

Detaching from unhealthy habits, addictions, ideas, thoughts, people, objects, behaviors can be freeing. It can also be frightening. "If I **detach,**" people sometimes ask, "what will be left? What will take the place of that person, place thing, habit, addiction, idea, thought, or emotion?" Change can be frightening. Our identity and sense of ourselves, positive or negative, is wrapped up in our attachments.

But **nonattachment,** like attachment, is quite a natural process. A few examples of natural detachment are separation, growth, space, death, difference, contrast, opposites. Detachment can be healthy or unhealthy, somewhere in-between, or neutral. Achieving nonattachment, like attachment, can be conscious or unconscious, in our awareness, or out of our awareness. **Awareness** is a key to understanding the self as well as to building or raising self-esteem. How do we become more aware?

Remember our mindfulness meditation exercises? Always bringing our **awareness** back to the present moment by focusing on the breath. In a sense, breathing is about attachment (taking the breath in on the inhale) and nonattachment (letting go of the breath on the exhale). Breathing is just one of many examples of natural processes of attachment and nonattachment. Can you think of others? Eating, crying, walking, sleeping...?

How do we detach ourselves from our ***unhealthy*** **attachments**? How do you let go? How do you "turn it over"? How do you separate? Suggestions:

1. **Be aware,** mindful, of attachment to others, thoughts, ideas, things, feelings—notice, watch, pay attention, observe (without judgment). **How do we do this**?
   a. Slow down (so you can see more clearly).
   b. Reduce the number of **mindless** activities in your day.
   c. Write down what you observe (in your day, in your thoughts, in your feelings) by journaling or listing.
2. **Ask yourself and others:**
   a. Is this (what I notice) healthy or unhealthy attachment?
   b. What purpose does it serve in my life? For example, does it fill my time? Does it help me cope? Does it affirm a core belief I have about myself? Does it give me an identity?
   c. Is it an addiction? (If so, can it really raise self-esteem? Remember self-esteem comes from within and an addiction involves a "substance" from outside of the self, and so the "fix" is temporary).
   d. Am I ready to let go, detach, separate from this unhealthy attachment? What ***resources*** do I need to tap for help? Physical resources, mental, emotional, spiritual resources (remember the medicine wheel).
3. **Create space in your mind and life for change, growth, and possibility.**
   a. Make healthy self-esteem a priority in your life.
   b. Imagine or visualize your self with positive self-esteem.
   c. Imagine or visualize an "inner guide" who can help you feel better about yourself. It's important that ***you*** choose the inner

guide—it could be a person, a symbol, an animal, a spiritual leader or image, for example.

d. Affirm your self with loving kindness.
e. Grieve losses (so you can move on).
f. Forgive yourself and others (so you can move on).

HOMEWORK EXERCISE

Develop a plan, using the "how to" suggestions above, for detaching from a person, thought, behavior, or object that represents an unhealthy attachment for you. Practice your plan every day.

REFERENCES

Ainsworth, M. D. (1973). The development of infant-mother attachment. In B. M. Caldwell & H. N. Ricciuti (Eds.), *Review of child development research.* Chicago: University of Chicago Press.

Ainsworth, M. D. (1979). Infant-mother attachment. *American Psychologist, 34,* 932–937.

Bacal, H. A., & Newman, K. M. (Eds.). (1990). *Theories of object relations: Bridges to self psychology.* New York: Columbia University Press.

Bowlby, J. (1969). *Attachment and Loss: Vol. 1. Attachment.* New York: Basic Books.

Bowlby, J. (1973). *Attachment and Loss: Vol. 2. Separation, anxiety, and anger.* New York: Basic Books.

Bowlby, J. (1979). *The making and breaking of affectional bonds.* London: Tavistock.

Bowlby, J. (1980). *Attachment and Loss: Vol. 3. Loss, Sadness, and depression.* London: Hogarth Press.

Bowlby, J. (1982). Attachment and loss: Retrospect and prospect. *Journal of American Orthopsychiatry Association, 52,* 666–668.

Brown, L. S., & Ballou, M. (Eds.). (1992). *Personality and psychopathology: Feminist appraisals.* New York: Guilford Press.

Carlson, E., & Sroufe, L. A. (1995). Contribution of attachment theory to developmental psychopathology. In D. Ciccetti & D. J. Cohen (Eds.), *Developmental psychopathology* (Vol. 1, pp. 581–617). New York: Wiley.

de Shazer, S. (1985). *Keys to solution in brief therapy.* New York: Norton.

Eliade, M. (1969). *Yoga: Immortality and freedom.* Princeton, NJ: Princeton University Press.

Epstein, M. (1995). *Thoughts without a thinker.* New York: Basic Books.

Ethical principles of psychologists and code of conduct. (1992). *American Psychologist, 47,* 1597–1611.

Farber, B. A., Lippert, R. A., & Nevas, D. B. (1995). The therapist as attachment figure. *Psychotherapy, 32* (2), 204–212.

Fontana, D. (1990a). Self-assertion and self-transcendence in Buddhist psychology. In J. Crook & D. Fontana (Eds.), *Space in mind: East-West psychology and contemporary Buddhism* (pp. 42–59). London: Element Books.

Fontana, D. (1990b). Self and mind in psychological counseling. In J. Crook & D. Fontana (Eds.), *Space in mind: East-West psychology and contemporary Buddhism* (pp. 205–212). London: Element Books.

Guidano, V. F., & Liotto, G. (1983). *Cognitive processes and emotional disorders.* New York: Guilford Press.

Gyatso, T., the Fourteenth Dalai Lama. (1992). *The meaning of life from a Buddhist perspective* (J. Hopkins, Trans. & Ed.). Boston: Wisdom Publications.

Gyatso, T., the Fourteenth Dalai Lama. (1997, June). Plenary sessions at a conference with Nobel Peace Laureates presented by Tibet House. *Peacemaking: The power of nonviolence.* San Francisco.

Holmes, J. (1993). *John Bowlby and attachment theory.* New York: Routledge Press.

Horowitz, M. J. (1988). *Introduction to psychodynamics: A new synthesis.* New York: Basic Books.

Joyce, E. (1997). *Through the eyes of Bodhisattvas (Awakening warriors): A qualitative study of Buddhist psychotherapists.* Unpublished master's thesis, Smith College School of Social Work, Northampton, MA.

Kabat-Zinn, J. (1990). *Full catastrophe living.* New York: Dell Publishing.

Kagan, J. (1996). Three pleasing ideas. *American Psychologist, 51*(9), 901–908.

Klein, M. (1952). Some theoretical conclusions regarding the emotional life of the infant. In M. Klein, P. Heimann, S. Isaacs, & J. Riviere (Eds.), *Developments in psycho-analysis* (pp. 198–236). London: Hogarth Press.

Kohut, H. (1977). *The restoration of the self.* New York: International University Press.

Kornfield, J. (1993). *A path with heart.* New York: Bantam Books.

Langer, E. J. (1989). *Mindfulness.* New York: Addison-Wesley.

Linehan, M. (1993). *Cognitive-behavioral treatment of borderline personality disorder.* New York: Guilford Press.

Low, J. (1990). Buddhist developmental psychology. In J. Crook & D. Fontana (Eds.), *Space in mind: East-West psychology and contemporary Buddhism* (pp. 115–122). London: Element Books.

Mahoney, M. J. (1990). *Human change processes.* New York: Basic Books.

Main, M., Kaplan, N., & Cassidy, J. (1985). Security in infancy, childhood and adulthood: A move to the level of representation. *In* Growing points of attachment theory and research. (I. Bretherton & E. Waters, Eds.), *Monographs of the Society for Research in Child Development, 50,* (1–2, Serial No. 209).

Mason, W. A. (1997). Discovering behavior. *American Psychologist, 52*(7), 713–720.

Maturana, H. R., & Varela, F. J. (1992). *The tree of knowledge: The biological roots of human understanding* (R. Paolucci, Trans.). Boston: Shambhala. (Original work published 1987)

Neimeyer, R. A. (1993). An appraisal of constructivist psychotherapies. *Journal of Clinical and Consulting Psychology, 61*(2), 221–234.

Rinzler, C., & Gordon, B. (1980). Buddhism and psychotherapy. In G. Epstein (Ed.), *Studies in non-deterministic psychology.* (Vol. 5, pp. 52–69). New York: Human Sciences Press.

Rosenbaum, R., & Dyckman, J. (1995). Integrating self and system: An empty intersection? *Family Process, 34,* 21–44.

Rubin, J. (1996). *Psychotherapy and Buddhism: Toward an integration.* New York: Plenum-Press.

Safran, J. D., & Segal, Z. V. (1990). *Interpersonal process in cognitive therapy.* New York: Basic Books.

Sroufe, L. A. (1979). Socioemotional development. In J. D. Osofsky (Ed.), *Handbook of infant development.* New York: Wiley.

Sroufe, L. A. (1996). *Emotional development.* New York: Cambridge University Press.

Stern, D. (1985). *The interpersonal world of the infant.* New York: Basic Books.

Varela, F. J., Thompson, E., & Rosch, E. (1995). *The embodied mind: Cognitive science and human experience.* Cambridge, MA: MIT Press.

Zukav, G. (1990). *The Seat of the soul.* New York: Simon & Schuster.

BIOGRAPHY

Bonnie Raber Wickes

Bonnie Raber Wickes is a psychotherapist in private practice in Anchorage, Alaska. She received her master's degree from the University of Alaska and is in her final year of the clinical doctoral program in psychology at the California Institute of Integral Studies in San Francisco. She has been an associate member of the American Psychological Association since 1986 and has been certified through the National Board for Certified Counselors since 1988. She is currently licensed as a psychological associate and a marriage and family therapist. Bonnie has been interested in attachment theory for years and presented her research on parental bonding and abuse at the American Psychological Convention in Los Angeles in 1985 and the International Congress on Child Abuse and Neglect in Sydney, Australia, in 1986. She is currently writing the results for her doctoral dissertation, a qualitative study of socialization process and development of nonviolence within the psyche. In the past few years she has found her own personal practice of yoga and meditation to be helpful in thinking about the attachment process.

5

PROBLEMATIC ATTACHMENT AND INTERRUPTED DEVELOPMENT OF RELATIONSHIPS: CAUSES, INTERVENTIONS, AND RESOURCES WITHIN A MILITARY ENVIRONMENT

RONALD G. BALLENGER

Clinical Child Psychologist
Educational and Developmental Intervention Services
Wuerzburg, Germany Region

INTRODUCTION

Relationships within a military environment are fraught with separations and losses. They cannot be avoided. Individuals within the military suffer losses and separations and continually cope with the anticipation of those in the future, whether planned or not. How to maintain a sense of safety and attachment in face of threat and how to maximize joy and endure sorrow are challenges with which military service members and their family members are constantly confronted.

Children within the military setting have significant issues with regard to separations, changes from one geographic area to another, having a variety of caregivers prior to entering a school setting, and being able to develop relationships in a foreign environment. The combination of these and other factors may induce fears of abandonment within a child with consequent depression, anxiety, and behavioral acting-out of these feeling through withdrawal and/or aggression. If the fears are not addressed the ability to develop trusting attachments to adults may be attenuated.

If the sense of neglect and/or abuse occur in the first 2 years of life due to frequent changes in caregivers or abuse/neglect by a specific caregiver, the likelihood of attachment problems arising is possible. Within the military environment, where a child may be reared by a sole parent or by two parents who are both military members, the possibility arises that a young child may be left in the care of a variety of caregivers for long periods of time, especially when the parent or parents are deployed to distant sites, lengthy training periods, or unaccompanied tours.

The lack of extended family to provide support to the sole parent creates a responsibility on that parent to develop trustworthy support networks to provide care, stimulation, and proper support for the child. The development of these networks can be a daunting task for a sole parent who may be feeling overwhelmed by one's own need for love, affirmation, and support. Absent the development of support for the child, though, especially for a child under the age of 2, issues of attachment disorder may appear. This is not to say that children over the age of 2 cannot develop attachments; they can but the process appears more problematic.

The need for affectionate ties between the parent(s) and the child is, of course, intuitively and experientially obvious. The child needs to feel secure, safe, loved, and nurtured in order to develop those feelings for others. The loving bond must be powerful enough to hold two individuals, at least, together even though they are separated by distance and long periods of time.

This emotional tie, this linkage of love, will serve to allow the child to understand the feelings of others and to respond in a loving, caring way to other individuals. This will provide the basis to develop friendships and intimate relationships in the future. The feeling that one is loved and attached to a loving parental figure allows the child to test one's own abilities both emotionally and intellectually. Children whom I have observed who develop this sense of attachment and of being loved are able to trust others and to develop a sense of trust in one's self and to feel reliant upon one's own resources.

RESOURCES WITHIN A MILITARY SETTING

Army Community Services, Social Work Services, and Family Advocacy Program

Child care resources are relatively constant across military installations. Most have Child Development Centers, Community Services, clinics for children with special needs, departments of psychiatry within the servicing hospital for a particular area, and other services particular to a certain area. This constancy of service provides a sense of stability and familiarity to service members and their family members when a permanent change of station (PCS) occurs, usually once every 3 to 5 years.

Usually, however, there is no mental health center dedicated to providing general mental health services; most agencies are established to provide particular services such as evaluating for and following up reported cases of sexual, physical, and emotional abuse. Social Work Services (SWS), the clinical component of Army Community Services (ACS), provide this service on most Army posts. There are counterpart agencies in the other military services.

The issue of child abuse, both physical and sexual, raises pertinent questions with regard to child–parent attachments. When an allegation of abuse is made to the Family Advocacy social worker in Social Work Services, an immediate report is developed and the family member who is indicated as the abuser is interviewed and may be separated from the family. The abuser, usually, but not always, is a male military member. He may be placed in separate quarters until the facts of the allegations can be evaluated and presented to the Case Review Committee (CRC) for a hearing and a vote can be taken as to whether to substantiate or unsubstantiate the allegations made.

The CRC is composed of various community individuals from a variety of agencies. The participation of these various agencies' individuals on the CRC is mandated by army regulation. The CRC listens to the presentation of the allegations and discusses, among themselves, with input from the Chair of the CRC, usually the Community Military Commander at the rank of Lieutenant Colonel or Full Colonel. If the Committee is satisfied that a full presentation of the facts has been provided, a secret vote is taken to determine if the facts, as presented, substantiate the allegation of abuse. The Chair is not a voting member unless the vote is tied.

The consequences of an affirmative vote may trigger a court martial, separation of the child from the family into a foster home, family therapy, individual therapy, or a combination of interventions as determined by the service member's commander, the community commander, and Family Advocacy personnel as well as the Staff Judge Advocate General's office. An affirmative vote in the CRC has far-reaching and powerful consequences to the military person involved and to the family.

The treatment and care of the child or children involved is the primary focus of all interventions made. The consequences of long-term sexual and/or physical abuse on a young child under the age of 5 may be to create behaviors somewhat similar to those observed via news reports in children reared in nonnuturing environments in the eastern part of Europe. However, the sexually abused child appears to respond much more avoidantly than these children do. This, of course, appears intuitively correct since these children needed to rely upon adult caregivers in order to survive whereas the abused child's welfare was enhanced by avoiding the abusing parent.

The attachment of the abused child to the abusing parent would, it would seem, be less secure and stable than to the nonabusing parent. However, the consequent behaviors resulting from the abuse may interfere with the attachments the child has with the nonabusing parent and, in fact, with other significant adults in the child's life. The issue that interferes with knowing whether the behavior exhibited by the child interferes with the attachment with the nonabusing parent is that it is difficult, if not impossible, to establish whether the child and nonabusing parent did, in fact, develop an attachment in the first place.

If it could be established whether what could be defined as a healthy attachment *did* occur in the first months and year of life, then one could determine if the behavior observed hence is a result of the abusive action of the abusing parent, which then reduces the attachment the child once had with the nonabusing parent.

Alcohol and Drug Abuse Prevention and Control Program (ADAPCP)

The Community Counseling Center (CCC) is established to provide evaluation and treatment for substance use, misuse, and abuse (Department of the Army, 1995). Some communities, depending upon the resources available and the desires of the commanding officer, will provide general mental health services through the Community Counseling Center even though, by military regulation, the centers do not have that mission. Other comparable agencies to provide evaluation and treatment for substance abuse are located in the other military services. The children of individuals who have substance abuse issues often have behavioral problems that are exhibited in other agencies within the military setting. These settings include, but are not limited to, the Child Development Center, Family Child Care (FCC) providers, Developmental Pre-School, and kindergarten.

If the military member has had a long history of substance misuse, it is probable that the effects of this misuse are shown, initially, within the family unit. Issues of nonattention to spouse and to children or verbal abuse to family members occur but are not reported at that time to Family Advocacy. The reluctance to report inappropriate behavior by a family member is, in the main, tied to fears, well founded, that the military member will suffer adverse consequences for the behavior.

The consequences could be a reduction in rank, nonjudicial punishment, court martial, removal from the military service, or even imprisonment if the violations are severe enough. Given the secrecy that the family maintains regarding the military members abusive use of substances, usually alcohol, the effects upon the children of the family are to behave in avoidant ways. Usually, the children will be socially disengaged and academically uninvolved such that their academic and social development is impaired.

The child or children involved have not, in most cases, been physically abused but they feel isolated and rejected by the substance-abusing parent. The non-substance-abusing parent, if that is the case, usually is so immobilized by the behavior of the spouse, that he or she does not have the wherewithal to provide emotional succor to the children. The children's relationship within the family may become frozen or static, initially, as they strive to maintain a sense of security, trust and protection within that unit. Trust is often a casualty of the dysfunctional dynamics within a family wherein one or both of the parents are substance abusers.

As the children within this type of family may not have developed a belief that adults will deliver on promises made or have abandoned that belief, a cynical view of adults develops. The child or children resist obeying adult directives or initiatives because experience has shown that there is no intrinsic or extrinsic reinforcement for having done so. Also, this behavior may be viewed by the child or children as a way to develop engagement with the parent(s) so that certain needs for attention and security can be met. Indeed, some children will eventually state, in their own way, that they do behave in aggressive ways to obtain a reaction and an interaction with the distant parent. The resistive behaviors are to elicit, however misconceived the thinking, some attachments with the parent or parents, should both of them be substance abusing.

The need for attachment, meaning the need to be close, emotionally and/or physically, to a person or, in some instances, an inanimate object, when one is scared, sick, or overwhelmed, is a fundamental need of children growing up in any environment. The need in a military environment is even more pronounced because of the transitory nature of the setting and the inherent danger of a caregiver being quickly, without warning, taken from the child. If the parent is, in fact, incapacitated by abuse of substances to provide the security the child needs, particularly during the first year to 2 years of life, the child will, very likely, act out fears of abandonment through aggressive acts and/or withdrawal.

Within the ADAPCP is located the Adolescent Substance Abuse Counseling Services (ASACS). The employee, the Adolescent Substance Abuse Counselor or ASAC, is a contracted employee, usually a social worker or masters-level counselor trained in substance abuse therapy with children and adolescents. This individual is located in a school setting, usually in the high school, and has the duty of providing evaluation and treatment to children who are either abusing or misusing a legal substance or are using an illegal substance. The ASAC is confronted with children with a variety of behavioral issues, the consumption of alcohol or use of psychotropic drugs being only a possible manifestation of the underlying attachment issues that have developed into disruptive behavioral disorders and/or substance-use disorders.

The ASAC will focus, as is the contracted mandate, primarily upon the issues of substance use and the issues of attachment or detachment from the family secondarily. The apparent low rate of success working with these, generally, adolescents reflects that other, more powerful forces regarding feelings of abandonment and depression are not fully addressed.

The usual goal for the ASAC is to stop the use of the substance that is under consideration. However, the more fundamental issue of healing or resolving attachments that were never made or were made and then were destroyed by subsequent parental–child behavior patterns is often times not addressed or if addressed, is not successful due to parental indifference or pathology.

Also, the therapists, while quite competent in the work that they are hired to evaluate and treat (substance issues), are not, usually, knowledgeable regarding attachment problems or interventions to address those problems. The focus of treatment remains upon abstinence from the use of the substance in question, which is, of course, a desired treatment goal, but the emotional wounds with regard to their parents of origin are not, typically, fully addressed. Again, this is often because the parents will not contract to be fully involved in the treatment process; rather preferring to blame the child for the unwanted behavior or to shame the child for what has been done, presumably, to the family image and name.

Exceptional Family Member Program (EFMP) and the Educational and Developmental Intervention Services (EDIS)

The Exceptional Family Member Program (EFMP) in the Army environment, predicated upon Army Regulation 608-75 (Department of the Army 1996) through its Educational and Developmental Intervention Service (EDIS), the clinical component of EFMP, provides evaluation and treatment for children from birth through 21 years of age. Children between the ages of 5 and 21 can be referred to the EDIS clinic by the Case Study Committees (CSC) of the various Department of Defense Dependent Schools (DoDDS).

The EFMP is mandated to care for children with medical, educationally related, and developmental difficulties if they have been found eligible for special education services. These are children who need the care of subspecialists. These are providers having skills and knowledge beyond that of a family practitioner or primary care provider. The subspecialists located in the EFMP include clinical child psychologists, speech and language therapists, occupational therapists, physical therapists, developmental pediatricians, and early childhood educators. Social workers are employed as program coordinators but also provide social work services.

A patient may be seen by a health care provider in a local medical treatment facility (MTF) and the decision may be made by the provider to enroll the patient into the EFMP by the use of a form designated as a 5862-R. The information on the form will be entered by the provider. This information includes a diagnosis, frequency of care, level of severity, whether inpatient or outpatient, and the type of care provider. The enrollment of the patient into the EFMP may have an impact on the next assignment of the military member but will not have an impact on the military member's career progression.

Enrollment in the EFMP is for all family members, not just children. A family member may enrolled if one has asthma or is suffering from muscular dystrophy or sclerosis, as examples. If services are not available to support the enrolled family member in the next gaining military unit, the military member and his family, usually, but not always, will be diverted to an assignment area that has facilities to support the family member(s) having special needs and requirements. The military service member may be directed to the gaining military unit and assignment and the family not permitted travel to that unit area. This event does not usually occur unless the military occupational specialty (MOS) is highly specialized and in short supply or if a particular officer is required for certain expertise that is possessed only by that officer.

The enrollment into EFMP does not mean that a child is eligible for special education enrollment, for special education services, or for medically related services (MRS) from the EDIS. This can be determined solely by referral from the Case Study Committees of the schools to the EDIS clinics. Enrollment in the EFMP does not equate to special education eligibility and services being provided by the EDIS providers. The services may have to be provided by other agencies in the community other than EDIS, unless the child has been found eligible by the CSC and an Individual Education Plan (IEP) developed, or is younger than age 3.

The services for a child who may be suffering from attachment-related issues probably will be provided by the Child and Adolescent Psychologist and/or Psychiatrist Services (CAPS) located within the Department of Psychiatry, which is located within the servicing military hospital for a particular region. Within the European command, there are three military hospitals with departments of psychology/psychiatry. These are located in Wuerzburg, Heidelberg, and Landstuhl, Germany. The child, 3 and under, may be referred back to the EDIS clinic for follow-up treatment by the Early Intervention specialists and the family provided, if it does not already have one, an Individual Family Support Plan (IFSP), which will delineate the services that the child and family need. This plan is developed in conjunction with the parent(s), providers, and other interested parties that are thought to be needed to enhance the child's development and progress.

The referral from the DoDDS CSC will request that a school-age child, enrolled in the DoDDS, be evaluated under specific categories pursuant to special education eligibility and placement. These categories include Autistic Disorder, Attention Deficit/Hyperactivity Disorder-Other Health Impaired and emotional impairment, meaning that if the child is found to be afflicted with a mental illness that will unlikely dissipate with medication and has an associated learning and/or cognitive problem, then the child may be found eligible by the CSC as eligible for special education placement and services.

Also, the child can be evaluated under the category referring to learning impairment, which requires that the child's achievement in math, reading, or language arts be near or below the 10th percentile or, for a child of above-average ability, near or below the 35th percentile. Another category is communication impairment, which includes voice disorders, fluency disorder, articulation disorder, and language disorder. The CSC will ask that specific questions be answered to the best of the evaluating staff's ability so that the CSC can fully appreciate the need for special education eligibility and services for the child under consideration.

The staffs of the EDIS include the Early Intervention specialists, noted earlier, who evaluate and treat children from birth to 3 years old. The children who typically present at the EDIS clinics have various developmental delays usually including language, motor, and behavioral issues in addition to feeding and sleeping concerns. Some of these disorders are, likely, a result of poorly formed attachments in the first months and year of life, particularly the behavioral and speech deficits.

OVERVIEW OF A COMPOSITE CASE

The developmental pediatrician serving that EDIS clinic often is asked to evaluate a child in preschool, kindergarten, or very early primary grades for medical issues by the referring Case Study Committee. Also, the Early Intervention specialists may refer a child under age 4 to the pediatrician for a medical work-up. This evaluation sometimes results in a referral from the pediatrician to a neurologist in the local servicing military hospital or to another hospital if the patient needs a particular subspecialist to evaluate the child, such as a pediatric neurologist or cardiologist. Also, there may be referrals to the clinical social worker for an evaluation of the home setting and the ability of the parents to provide the support the child may need and to the clinical child psychologist for an evaluation of the parental unit and the relationship between the parental unit and the child.

Sometimes these evaluations reveal that the inability of the child to regulate one's feelings and emotions are likely due to a combination of factors impacting upon the parent(s) and the child. These factors may include a series of separations immediately subsequent to birth that have interfered with the mother's capacity to attend to the child consistently. These separations are usually because of school requirements for the active-duty military person. These trainings may require that the service member return to the continental United States (CONUS) for lengthy periods. The mother may then return to her duty station and begin to attend to her child only to receive orders to report for an extended field training exercise (FTX). Once that FTX is completed, the service member may then be deployed to any number of sites in the world for a 3- to 6-month stint.

The child has not suffered physical or sexual abuse and has had one's physical needs attended to by surrogate caregivers, usually several caregivers. But the series of attachment breaks over the first several months, even the first year or more, of life appear to create within the child a reduced ability to develop a sense of security and trust. Attachments to the mother are seriously attenuated.

Children seen at a later stage of development, approximately ages 4 or 5, as they enter into preschool activities, exhibit behaviors that are usually aggressive, show underdeveloped language skills, and are low on standardized cognitive tests such as the WISC III. Subscales measuring comprehension and social awareness seem to have some sensitivity to the impact of early attachment breaks even though the breaks were not characterized by abuse or severe neglect.

For example, some children born immediately before, during, or immediately after the conflict in the Arabian Gulf in 1990–1991, when seen in 1995–1996, exhibited behavior that appeared to be a result of inconsistent patterns of attachment. Some children, upon birth or soon thereafter, were returned to CONUS to be with immediate family, usually grandparents, while the military member(s) were deployed to the Gulf or were preparing to deploy. Others were returned to CONUS and then were returned to outside the continental United States (OCONUS) when the caregivers in the CONUS setting found that they could not provide the level of care needed for the child. The child, upon returning to OCONUS, was placed with surrogate families until the return of the birth parent or parents, sometimes weeks or months after the child's return.

Some children, experiencing the above or similar circumstances, when returned to the birth parents, exhibited very problematic behaviors. Some were overly demanding and showed some speech delays. When left in child care agencies, screaming and fighting ensued. Threats toward the parents or the caregiver could be made if demands were not immediately carried out. Oppositional and defiant behavior was common.

What effect the varied deployments to the Balkans will have upon infants and very young children is not, at this time, known. Given the implementation of early intervention services in 1995 and the lessons learned from the Gulf War that family care plans must be viable and in place, it may be that the effects of attachment breaks can be ameliorated.

Interventions that did have useful outcomes with children who apparently were suffering from some type of attachment disorder included interventions that were behaviorally predicated. The use of insight counseling, while appealing to a degree because of the superficial charm of some of the children during the counseling hour, did not appear to result in any significant changes in behavior. Also, those children who had a sense of firm values based upon learnings from a stable family member, church groups, or civic groups or some other stable source seemed to be less overwhelmed by the loss of a deployed parent and did not exhibit the levels of internalized or externalized behavior that children who did not have such a background.

With children ages 5 and above, the development and implementation of firm boundaries with clear and certain consequences in the school and home settings produced some reduction in oppositional/defiant behavior. The use of reinforcers such as being allowed more time with a desired person, attending specific events, or being allowed to have particular items, such as compact discs, did have the result of the child producing more academically and developing more positive relationships with adults and children.

The child seemed to understand that even though there were powerful feelings of being abandoned or unloved and that even if she acted these feelings out that there would be caregivers, therapists, and teachers who would respond to him or her in a caring, loving manner even though some consequence might be imposed such as a loss of privileges. Apparently, a sense of trust did eventually develop because the child began to understand that he or she could not control the adults involved. He or she began to feel safe and trusting within that school and home environment.

Interventions, then, that do not include holding of the child seem to have some efficacy although, of course, it may be argued that the behavior being observed was not, in fact, an attachment disorder but rather behavior stemming from other factors. The observation of a field intervention does create a basis for controlled studies with regard to children within an academic setting who have suffered apparent attachment breaks during their early years due to work-related issues of parents who are military service members.

Some children who can have their feelings accessed through talking therapy do appear to be able to have their perceptions changed and, consequently, their feelings. These feelings, once changed to those of trust and security, can be projected onto their parents and caregivers rather than the feelings of rage induced by the feeling of abandonment. The talking therapy that has, seemingly, from a clinical perspective, had an impact is characterized by intense focus, good eye contact, which must be maintained by the therapist, quick, incisive, verbal interchanges, frequent touching of the knee and upper arms, if the patient will tolerate it, and a sense of humor couched with warmth.

The child often will present in the sessions with a hat or cap pulled down over the eyes, a hostile sullenness, and, at times, a barely concealed aggressiveness. The child is requested to maintain eye contact and if that responsibility is ignored the therapist must work to maintain that contact. If the eye contact is not kept and if the verbal interchange, usually reflecting the child's apparent state of feeling and thought with additional statements regarding the child's relationship with current caregivers and memories of earlier relationships with current or prior caregivers, is not maintained, then the therapy session does not accomplish the goal of developing an emotional contact between the child and the therapist and treatment cannot proceed.

Much of the dialogue in the initial sessions is unbalanced. The therapist has the task of developing a monologue, practically, describing the child's state of mind, possible memories, emotionality and so forth. This necessitates the therapist having a fairly good grasp of the child's history and current experiences. Even with this knowledge, the implementation of the initial sessions takes great energy and not a little creativity on the part of the therapist to engage the child sufficiently so that the child in subsequent sessions will begin to provide information and data regarding prior memories and current behaviors and feelings.

The development of a good interview with the most prominent, knowledgeable caregiver of the child can provide useful material to speak to the child about when the actual session begins. This will also highlight how motivated the caregiver and/or parent(s) are to be involved in the assessment and treatment of the child.

Sometimes, the military parent, who is so overwhelmed with the military duties required, develops the belief that the child's therapy is solely the responsibility of the therapist and that parental involvement is not needed. This belief, if it is uncovered in the initial interviews, needs to be disabused immediately. Follow-up interventions outside of the consultation room on the part of the parents will help to create attachments between the parent and child such that the child can function more fully.

The child experiences the parent's behavior as insensitive to the his or her emotional state. In an older child of 4 or 5 years of age who is referred by the child development center or the kindergarten teacher, the presenting behaviors are often oppositional/defiant or uninvolvement with the classroom activities. The child does not appear to be in touch with others in the class either emotionally or physically. The avoidant behavior is usually the most noticed behavior, initially. The oppositional/defiant behavior is exhibited when the child is directed to contribute to a classroom discussion, follow classroom procedure, or to discuss or comment about one's behavior in a specific situation. The oppositional behavior is usually the behavior that triggers a referral. The child appears to have little capacity to speak about one's self or to connect feelings with actions.

Other children, who are referred for evaluations pursuant to special education eligibility, may present as rather anxious children. They tend to be worried about a variety of issues such as whether another child perceived as a "bully" will beat them up after school; if, after hearing a report that a possible snow storm is approaching the school will be open the next day; or will their parents keep their siblings from using their new video game.

The child seems to be worried constantly about how dependable and predictable other people are. Interviews with parents may reveal that the parent(s) has unresolved issues from his or her past that intrude, often, into their present-day life. They are so consumed with these past episodes in their life that they are not available to consistently provide emotional support to their child or children.

The child usually will receive diagnoses of generalized anxiety disorder and/or dysthymia. Sometimes there are attendant thought-disorders that are, seemingly, a function of the feelings of anxiety. Language skills appear to be attenuated, especially when attempting to express an abstract thought—something not connected to the immediate environment—with visual detail.

Talking therapy sometimes has an efficacy in that the child, usually those past age 5, will modulate behavior and emotion. Again, the therapist has to address what appears to be the underlying cognitions of the child and surface those thoughts by expressing them as possibilities as to what the child may be thinking and feeling. If, by this tracking and interactive process, the therapist is relatively accurate, the child can begin to feel some trust in the therapist and, in turn, to relate to the therapist what is actually being felt or thought and reduce the guessing on the part of the therapist as to what may be happening internally within the patient. The therapist must stay engaged with the patient and not just react to the patient but maintain an active involvement with the patient.

The therapist should comment on what can most easily be confirmed as being true, especially in the first contacts with the patient. This will create a positive response set within the patient such that when an inaccurate comment is made, the child will accept the comment and indicate to the therapist that the comment was incorrect and provide information as to what was actually being felt or thought.

Other children in a military setting may be referred by the CSC, for evaluation pursuant to special education eligibility subsequent to having been evaluated by other providers, such as pediatricians, neurologists, social workers, and psychiatrists. These children present with self-destructive thoughts and, in some cases, actions. These self-destructive acts have, sometimes, not been documented in the medical record because they occurred prior to the family entering the military or the record was stripped of these entries during a prior PCS move.

These children present the most worrying behaviors of any. Some of these patients appear to be in a kind of trance in the school setting and/or make odd, weird statements that are unsettling to the other children and to the school personnel. These behaviors and statements may be subtly or overtly aggressive rather than internalizing, which may lead one to surmise the child is angry toward others but, in fact, the patient is more depressed and unloving of him or herself than angry toward others. Children who present these behaviors are at very high risk.

These children have diagnoses of Attention Deficit/Hyperactivity Disorder, learning incapacitations, especially with poor working memory, and Opposition-Defiant Disorder or Conduct Disorder. They have difficulty in attempting to express themselves with coherency. Work with a patient exhibiting these behaviors usually will be ended because of behaviors of one or both of the parents in the community or because of the behavior of the child. If the child and/or parents exhibit behavior that is viewed by the military command as being an embarrassment to the command and to the United States government, the family may be immediately returned to CONUS and the military member separated from the military service (R. Flevotomas, personal communication, 1998).

Thus, these families often do not remain long enough within the military service to be accorded the services that are necessary in order for the attachment issues to be properly addressed both for the child and the parents. The resources needed to address these particular issues may not be available in OCONUS and the family may be returned to CONUS for further treatment if it is felt by the command that the military member can continue to contribute and function in one's military occupational specialty.

If the command does return the child and/or the family to CONUS, special attention may be given to locating the child near a military hospital that may be able to provide treatment to the child and the family that will reduce the observed behaviors. This treatment may be provided in an inpatient facility located in the military hospital or a day care facility. The treatment may include, but not be limited to, group treatment wherein children with a variety of diagnoses may interact in group psychodynamics and, too, work with structured group exercises. Also, individual counseling may be accorded the child as well as family sessions.

If the child and a parent, but not the service member, have been returned to CONUS to receive treatment in a military facility or other appropriate facility, and the child and parent are then returned to OCONUS with recommendations for follow-up treatment upon completion of the treatment in CONUS, the receiving unit will have to develop an appropriate treatment plan for the child and family. Interventions that seem to have had the most efficacious results with children exhibiting conduct problems resulting from attachment disorders formed in early life (due to separations and attachment breaks with the primary caregiver, not children who have suffered severe physical and/or sexual abuse) are implementation of parent retraining in appropriate child management procedures.

Social skills classes initiated by the school counselor are often useful in helping the child develop better peer relations. Also, if the child has qualified for special education services under one of the categories alluded to earlier, and an IEP developed, then the child may receive modifications within the classroom setting that may have the effect of helping the child to succeed within that domain. Often children exhibiting behavior problems also have great difficulty in formulating thoughts and expressing them in writing. If those requirements are modified and one is permitted to, perhaps, tape record thoughts and responses to classroom demands or has but to write only a portion of the requirement and record the remainder, then the behavior may improve as well as there possibly being an increase in academic progress.

The development of a parent training program within a military context is a daunting challenge. The military parent has a typical duty day from 0730 to 1700. Many have duty days that go beyond those times, especially when preparing for a major deployment to an area such as Kosovo or the Middle East, or for a major FTX, or for inspection from higher headquarters. These events can consume up to 12 to 14 hours per day for several weeks. Thus, by necessity, unless the military parent's commander will release that parent to attend the parent training class, the nonmilitary parent, if that is the case, will have to attend the parent training class and relay the training to the military spouse at home. This, of course, is not an ideal way to train parents in how to manage a child who has serious conduct problems.

If the child has committed offenses that have come to the attention of the Community Civilian Misconduct Authority (CCMA), usually the Deputy Commander of the post or the Executive Officer, communication between the individual developing the parent training program and the CCMA can usually result in the military parent being released from duty to attend the training sessions. The relationship between the individual providing the training and the CCMA will, many times, determine whether the military parent will be released by the unit commander.

The usual training provided to parents who have behaviorally disordered children, under the age of 10, consists of teaching how to consistently apply appropriate consequences for undesirable behavior and powerful reinforcers for desired behavior. Although this is a fairly simple concept, teaching parents to employ it is not. The assessment procedures to evaluate parental patterns between each parent and of each parent in relationship to the child takes a substantial period of time; to teach alternative behaviors takes an equally substantial amount of time.

Working with a child who is avoidant can be incredibly annoying and frustrating to the parent(s) involved as well as to the therapist. The caregiver or therapist has to figure out some way to entice the patient, especially a patient over the age of 5, to engage. This necessitates, as stated earlier, the employment of skills that will put the therapist in some synchronicity with the patient's internal and behavioral state. The acceptance of the behaviors and, in particular, the feelings presented by the patient in the therapy session will, it is hoped, eventually help the patient to adapt to and accept his or her own feelings. The patients' ability to accurately name the feelings and needs they are experiencing helps the relationship to develop. As this is done, it is hoped and expected that a more loving, trusting relationship can be created between the patient and the care provider, whether that be a therapist, parent, or surrogate parent.

The overall training period, including practice, can be accomplished over several weeks. The effects of the training are usually noticed within 4 to 6 weeks of implementation of the learned material. Whether these children are, in fact, suffering from attachment disorders which result in oppositional-defiant or conduct disorders is very difficult to determine. Many of the children who have these disorders have experienced breaks from their primary caregiver, the military service member, in the early part of their life. Whether these breaks were the precipitating causes for the subsequent behavior disorders is impossible to say with certainty, but they do appear to be significantly implicated from a clinical perspective.

SUMMARY

This chapter has addressed the possible development of attachment disorders within a military environment. These disorders may be manifested because of certain unique aspects found within a military setting. Attachment issues may arise because of excessive separations from the child by a parent due to military demands. These demands may include the necessity to be absent from the child for long periods due to training or educational requirements that the service member must have to maintain military readiness and effectiveness. Also, military inspections by higher headquarters may consume enormous amounts of time and energy on the part of the military service member to ensure that the unit meets appropriate combat standards. Field training exercises that may take the service member far from home for long periods of time may also contribute to the sense of detachment on the part of the child. Additionally, deployments to combat zones or zones of peacekeeping activities for long periods, sometimes up to 6 to 9 months, contributes to the sense of abandonment or detachment on the part of the child.

Also, the child may be living in a home where one or both of the parents have problems surrounding the use of chemical substances and the child's needs are not appropriately met and inappropriate boundaries are established. The child may respond to these situations with avoidance, aggressiveness, indifference, or any combination of behaviors. Interventions can include a variety of steps from mild reminders to the child, to intensive outpatient treatment, to inpatient treatment with parent retraining as to how to appropriately manage a child who has had severe attachment breaks. The child, even as old as age 10, can receive interventions that will ameliorate the undesired behavior to the extent that the child is able to function relatively well in typical social settings. This is an outcome that is devoutly to be wished for since the alternative is ongoing conflict with societal rules and regulations that may result in the child being unable to lead a fairly happy, rewarding life.

ACKNOWLEDGMENTS

The views presented in this chapter are those of the author and do not necessarily reflect those of the Department of Defense or the Department of the Army.

REFERENCES

Department of the Army. (1995). *Army Regulation 600-85* Change 2*. Washington, DC: Alcohol and Drug Abuse Prevention and Control Program.

Department of the Army. (1996). *Army Regulation 608-75*. Washington, DC: Exceptional Family Member Program.

R. Fleuotomas, personal communication, 1998.

BIOGRAPHY

Ronald G. Ballenger

Ronald G. Ballenger, A.B., M.S., Ph.D., earned his Ph.D. from Ball State University and A.B. and masters from Indiana University. He was for 2 years the staff clinical psychologist at Northeastern Center in Kendallville, Indiana. Prior to that, for 4 years, he was Professor of Counseling Psychology in the Ball State University European Counseling Program and the Boston University Overseas Program. He was President of the European Branch-American Counseling Association and, later, EB-ACA Treasurer. He was selected the Outstanding Member of that organization in 1988. For over 6 years, Dr. Ballenger was the Clinical Director of the Community Counseling Centers in Grafenwoehr and Munich, Germany, providing assessment and treatment in the area of alcohol abuse and drug use. He has presented numerous workshops on various topics to military and civilian groups throughout Europe. Dr. Ballenger, currently, is one of two, Clinical

Psychologists with the Educational and Developmental Intervention Services, the clinical component of the Exceptional Family Member Program in the Wuerzburg, Germany Bavarian Region. He recently contributed a chapter entitled "Counseling/Clinical Psychologist's Role in the Military" in a textbook entitled *Military Psychology: An Introduction* published by Simon and Schuster. His research interests are in the area of harm reduction for alcohol use and in the development of empathy in psychological caregivers. To contact, write: 235th BSB, CMR 463, Box 511, Army Post Office AE 09177, or phone/fax: 011-49-9828-1337, or E-mail: 106407.3130@compuserve.com.

6

PERMANENCY PLANNING AND ATTACHMENT: A GUIDE FOR AGENCY PRACTICE

DIANE SABATINE STEWARD
KATHRYN ROSE O'DAY

Children's Home Society of Florida
401 NE Fourth Street
Ft. Lauderdale, Florida

Chrissy did not enter the child welfare system by choice. She was drafted for her participation by various adults: her biological caregiver, who intentionally or unintentionally abused, abandoned, or neglected her, and other adults, who intervened with every intention of removing her from harm's way and providing conditions in which she could flourish. Chrissy herself hopes that finally now she can begin to have a life like she imagines other children do, though secretly she wonders if she really deserves it. After all, she must have done something wrong to make her mother act so badly. Now she can't even live at home anymore.

Once in the child welfare system, the number of adults who affect Chrissy's life increases exponentially, and each one frequently has a different perspective on which decisions need to be made "in Chrissy's best interests." These adults may include Chrissy's protective investigator, her juvenile police officer, her social worker, her Guardian ad Litem, her judge, her therapist, and others. Often one or both of Chrissy's biological parents will also be included. Chrissy's situation is muddled, because her parents are among those who are marginally capable and yet love her to the best of their ability. At times they can demonstrate some effort towards reunification, but consistently fall far short of adequately providing for all of Chrissy's major needs.

The result of this is that Chrissy may linger long in alternate care, waiting day by day for all the adults involved to agree to a plan for her care which will meet her needs and, most importantly, provide a permanent family environment which is not subject to turmoil and disruption. At times (more often than the authors would like to see) the system fails, a stable and lasting solution is not found, and a child like Chrissy waits and grows to chronological and biological maturity in a state of limbo. Can Chrissy attain emotional maturity under these conditions? Can she grow into a person who can form appropriate productive relationships and break the intergenerational pattern which frequently is seen in families in the child welfare system? The way in which we, as professionals, foster parents, and concerned advocates, care for Chrissy may make all the difference in the answer to these questions.

Chrissy's success or failure in her future life may well depend upon an attribute which we know as "attachment." Attachment can be briefly defined as the capacity for emotional security, closeness, and autonomy which develops in a child who has experienced during early life a predictable pattern of warmth, sensitivity, responsiveness, and dependability from a significant caregiver, usually the mother (Karen, 1994). Secure attachment is thought to be necessary for emotional health and to form the bedrock for all productive and successful relationships in childhood and later on in adulthood. Attachment theory has been an important influence in the debate regarding the role of "nature and nurture" in behavior, child development, and psychotherapy since it's conceptualization in the 1960s.

Chrissy is at great risk for problems with attachment. First, she requires the services of the child welfare system because her home life, and her parents' ability to care for her, were far less than ideal; in fact, they were worse than inadequate. Parents who are emotionally or chronologically immature, substance abusers, chronically ill, emotionally incapacitated, or who lack the fundamental human empathy to see a child as a separate human being, place a child at risk for attachment problems. Under any of these circumstances, the basic conditions of predictable and caring responsiveness from a caregiver are severely compromised. Second is the simple fact that Chrissy is in the child welfare system at all. Our system of care for dependent children has only recently begun to understand the causes and ramifications of attachment problems and has just formed a nascent awareness that the system itself may unwittingly contribute to these difficulties.

There are a multitude of issues regarding permanency planning for children who enter the foster care system. Many of these issues have been and continue to be widely discussed and debated among those of us who work in this arena. The authors continue to be concerned about all of the various facets of permanency planning. This chapter, however, focuses specifically on the consideration of promoting and preserving healthy attachment as a child makes his or her way through the foster care system.

Expanding demands on state child welfare agencies, coupled with shrinking resources, have created a highly overburdened system which has been increasingly unable to meet the need for safe and supportive foster care. Our private agency, traditionally involved in family preservation casework and infant adoption, has in recent years found itself drawn into providing supported foster care as our state agency has become less and less able to do so effectively. The observations presented here have evolved from our attempt to provide a more responsive and sensitive experience for the children who have been entrusted to our care.

There was a time when the lack of secure attachment was seen as a relatively rare event, observed mostly in hospital settings with premature or ill infants separated from their mothers for prolonged periods during the early crucial weeks and months of life. Over the past 20–25 years, however, a number of factors have combined to produce many more children coming into care without the necessary foundation of a secure attachment to help them cope with the losses they will experience. These children then encounter a system that further undermines their fragile connections and limited emotional stability.

This chapter examines four critical aspects of substitute care that can have an important influence on supporting or fostering healthy attachment in dependent children in the child welfare system. These areas are assessment, recruitment of substitute families, training and support of foster and adoptive families, and special issues. In each area of focus, we present a case from our work with children and families that illustrates potential problems, strategies, and solutions. All names of clients and some identifying information have been changed to shield the confidentiality of the child and family. Finally, we present some structural considerations for the agency or organization that is aware that its policies and procedures can have a profound impact on a child's experience with and capacity for attachment as he or she moves through the system of care.

We assume the reader of this chapter has familiarity with the concept of attachment and an understanding of the conditions that foster secure bonding between child and caregiver early in life. For the reader who is new to the topic, we suggest the excellent booklet by Elizabeth Randolph (1997) as a quick and easy way to gain a working knowledge of the subject. This booklet is also a handy resource for foster parents and other substitute caregivers who may need to be educated in the workings of attachment in order to cope effectively with the children in their care. The reference list at the end of chapter also serves the reader who wishes to pursue the theme of attachment in more depth.

ASSESSMENT

Comprehensive assessment is the cornerstone of effective permanency planning, particularly as it relates to facilitating the development or strengthening of attachment bonds. There is less potential for error in making appropriate placement and treatment decisions if the initial assessment has been thorough. The case of Hannah C. illustrates the importance of this point:

> Hannah came to the attention of child welfare at two weeks old, when she sustained two broken bones in her forearm. Her parents indicated that she had fallen and, since their story was somewhat plausible, she was allowed to remain with them. Two weeks later, Hannah was brought back to the hospital with mouth and face injuries, as well as pneumonia. Her mother appeared to be very inexperienced as a caregiver and detached from Hannah. Her father was angry and aggressive, clearly dominating his young wife. At this point Hannah was removed from the home and placed with the N. family. Her foster parents noted that she was quiet and appeared to be content to lie in her crib without much, if any, interaction with or stimulation from them.
>
> As time went on, Hannah slowly began to respond to the N. family; however, at approximately seven or eight months old, she was returned to her biological family, despite the indication of some serious parental deficits. Within a month, Hannah was back at the hospital with a broken pelvis. Upon release from the hospital, she was placed in an emergency shelter. The shelter was not equipped to deal with an infant in a body cast, so she was subsequently transferred to a specialized infant shelter. Once she was stabilized, Hannah was sent to live with relatives in another state, where she remained for approximately 18 months. When the relatives could no longer care for Hannah, she was sent back to her home state.
>
> It was indeed fortunate that the child welfare system at least saw fit to again place Hannah with the N. family, who welcomed a drastically different child back into their home. Now two and a half years old, Hannah had suffered serious physical abuse and had been moved six times. She displayed severe tantrums, "trashed" her room, put holes in the wall, and physically hurt both family pets and her foster sibling. Her behavior was distinctly unmanageable, even for these experienced foster parents.

Had the initial assessment of Hannah and her biological family been more thorough, several crucial factors might have come to light regarding the low level of functioning in her parents and the limited possibilities for reunification. Had Hannah's quiet and withdrawn behavior been noted and recognized as a potential attachment problem during her first placement with the N. family, subsequent decisions about placement might have either limited additional moves or taken special care to work on attachment issues with her.

Much unnecessary pain and placement failure can be prevented by attending to attachment in assessing children as they enter the child welfare system. Biological families must also be assessed for attachment issues, since the parent who is attachment impaired cannot give the child what that parent does not possess: the capacity to attach securely. Special attention must be given to these intergenerational patterns of attachment, as they can have immediate and critical implications for reunification decisions and permanency planning.

A comprehensive assessment should be conducted by a clinician with a good working knowledge of attachment theory, the signs of insecure attachment, and signs of attachment disorder. As mentioned earlier, a lack of secure attachment has traditionally been thought of as rare and thus may be easily overlooked in formulating a diagnostic impression. Many clinicians are familiar only with the DSM-IV description of Reactive Attachment Disorder of Early Infancy and Childhood. (American Psychiatric Association, 1994). It has been our experience that many of the older children we see in the child welfare system who have been given diagnoses such as ADHD, Conduct Disorder, and Oppositional Defiant Disorder often have undiagnosed attachment issues as the foundation of these behaviors. It is easy to overlook the profound insecurity of poor attachment when confronted with an "in-your-face" defiant child. A clinician with experience in working with abused and/or neglected children may be more adept at recognizing some of these subtle distinctions.

It must be acknowledged that it is difficult (but not impossible) to do appropriate and accurate screening during the initial removal of a child from his or her home. Frequently, parents are hostile and suspicious and children are fearful and withdrawn. It is incumbent on the clinician to communicate empathy, interest, and a genuine desire to learn from the family any information that will help the agency effectively care for their child. Extensive information regarding the prenatal and infant/toddler years of the child as well as the parent's impressions of his or her own childhood, and the affect and intonation with which they speak about these issues, provide beginning clues to the bond that flows between parent and child. This information should then be augmented in the early days and weeks by detailed information about the child's affect and behavior from shelter personnel and/or foster parents as well as direct observation of the parent/child relationship during family visits. Thorough psychosocial assessment of both child and family is critical to initiating appropriate placement and treatment as early as possible.

RECRUITMENT OF FOSTER/ADOPTIVE FAMILIES

The recruitment of foster and adoptive parents is often a daunting task. Demands for family-based care are high and increasing every year. Dual income families reduce the number of stay-at-home parents available to foster children. Often, agencies are faced with such a pressing need to find foster placements that any willing participants who can pass background screening and have appropriate housing are accepted. Considerations such as the emotional resilience of potential parents and the ability to deal with problem behaviors often take a back seat in an environment which is driven by a shortage of placement options in crisis proportions.

The challenge of recruiting appropriate foster families is further complicated by the fact that children entering foster care today suffer from more problems and have needs of a more serious nature than ever before. A number of factors, such as increased drug and alcohol abuse, HIV/AIDS, childhood poverty, and single-parent families have combined to produce a population of foster children who need far more than routine substitute care.

Although it is difficult to accurately pinpoint the prevalence of substance abuse among parents, data from the National Institute on Drug Abuse (NIDA) Household Survey for 1988 indicates that young adults aged 18–25 comprise the population of heaviest alcohol and drug users. At least 10% of pregnant women in several research populations was found to be using cocaine, and the use of marijuana in pregnancy was estimated to range between 14 and 32% (Ewing, 1993). Often the child's birth is an added stress which has resulted from an untimely pregnancy, as illustrated by a sample of 135 women identified with alcohol- and drug-abuse problems at the time of delivery. Ninety-seven percent of these women reported that the pregnancy was "unplanned" (Ewing, 1993). The relationship between alcohol and drug abuse and child abuse and neglect is consistent and supported by numerous studies. Parental substance abuse is closely associated with various levels of child neglect, ranging from inadequate supervision to a complete failure to provide for basic needs (Strassner, 1989). Children from alcohol- and drug-abusing families are also more likely to suffer fetal distress, Fetal Alcohol Syndrome, physical abuse, and sexual abuse (Gonzales-Ramos & Goldstein, 1989).

Pediatric AIDS also presents a challenge for foster care. It is estimated that about 1.5 newborns per 1,000 are born to HIV-infected women in this country—about 2,000 per year. Current estimates of the total number of HIV-infected children in this country range between 10,000 and 20,000 (Boland & Oleske, 1995). Due to the health status of their mothers, many of these children, with their myriad medical, emotional, and developmental challenges, will require substitute care either with a relative or in foster care.

Childhood poverty and single parenthood (which are closely associated with one another) have combined to produce children whom one author describes as "virtually unsupervised" (Lindsey, 1994, p. 72). According to the United States Census Bureau, about 2.1 million children younger than age 13 have no adult supervision both before and after school. Many of these children live in single-parent households, as the number of children involved in divorce has tripled, even while the overall child population has not changed in size. Many of these families live below the poverty line, as single mothers and their children comprise the largest group of people living at this income level (Lindsey, 1994).

Damaged biological parents often cannot help but produce damaged children. For example, a parent who has survived traumatic conditions such as childhood sexual abuse, physical abuse, loss of a parent, poverty, domestic violence, or chronic physical or mental illness may ultimately be a parent who is simply unable to respond appropriately and effectively to a child's needs. Another example is the adolescent parent who, due to the demands of completing his or her own developmental tasks, is unavailable (emotionally, physically, or both) to an infant and that infant's demands upon the parent. At times, these factors can combine to produce a parent and a child who together are seriously at risk. Foster children who become parents often may fall into one or both of these categories. This is not true in all cases, and is certainly not inevitable, but should be noted as a significant risk factor. Workers will often find that if they remain in the field long enough, they will work with several generations of the same family.

The effects of poverty, HIV infection, substance abuse, single parenthood, and emotionally damaged parents combine to produce a population of foster children who have often survived severe neglect and dangerous environments where drug-dealing, prostitution, and violence are the norm. The confluence of all of these factors has meant that many times children who manage to survive infancy will still have to negotiate a myriad of detrimental experiences: exposure to domestic violence; emotional, physical, and sexual abuse; neglect; developmental and emotional delays; behavioral difficulties; and insecure attachment. The challenge these children present to their foster parents as they arrive in out-of-home care is enormous.

Given these scenarios, the choice of foster and adoptive parents becomes even more crucial. They literally have to single-handedly fill a number of critical gaps in reparenting the child and understanding the discrepancy between the child's chronological age and his or her developmental abilities and emotional age. Foster and adoptive parents must have not only good basic parenting skills, but also specialized skills that allow them to respond appropriately to the child's emotional age rather than to his or her chronological age. At the same time, they must be able to give the child permission, encouragement, and guidance to accomplish the necessary developmental tasks. Woven throughout these skills must be the ability to provide the attunement experiences which will help the child develop and/or maintain secure attachments. Most parents can easily relate to the attunement activities one does with an infant (touching, cooing, eye contact, holding, smiling, talking, rocking, singing, etc.). Unfortunately, many may not understand that they must find ways to have that same type of "connecting" experience with an angry or withdrawn child who never had those important interactions as an infant. Brief touches and eye contact, playing games together, singing the child a lullaby, rocking, and giving the child a bottle are some potential attunement activities which can begin to provide the child a sense of belonging and attachment. The ability to expand on and adapt early attunement activities for an older child is a key feature of repairing the insecure attachment of many foster children (Hughes, 1997).

The ability to demonstrate empathy is an important corollary to providing attunement experiences. By using expressions of genuine empathy, parents model understanding and acceptance, establish a positive style of communication, and convey that the child's feelings are valuable. In addition, a response of "It sounds like you're feeling really angry at me right now . . ." helps the child to focus on his or her own feelings without defensiveness and encourages the child to modify his or her affect and behavior in future encounters. For children who have not experienced empathy in their families of origin, these exchanges may be their first opportunity to understand the appropriate give and take of functional family life (Hughes, 1997). In providing attunement and empathy, parents are assuring that their emotional responses match or complement the affect and behavior of the child. This task may become Herculean if the child is significantly attachment disordered; however, without experiencing attunement and empathy, it is unlikely that the child will be able to move forward and complete the appropriate developmental stages. It is, therefore, imperative that we recruit families who have an innate understanding of these concepts and are able to adapt their attunement and empathy skills to meet the needs of our special children.

If foster or adoptive parents do not have a strong attachment base themselves, it becomes difficult for them to parent children with attachment problems. Children with attachment disorders have an uncanny ability to tap into a parent's worst insecurities and control issues. If the parent did not have appropriate attunement experiences as a child, it becomes difficult to provide those same experiences to his or her foster/adoptive child. Parents must be very centered, reacting with firm but loving responses to difficult and challenging behaviors. They must be able (as much as possible) to resist interpreting the child's misbehavior as personal rejection. They must be able to love the child without expecting to be immediately loved in return and without becoming ego-involved in making the child fulfill the parents' needs. This is unquestionably a high ideal, which many of us fail to attain even as biological parents.

One good means of assessing potential foster and adoptive parents who have already parented is by looking at their relationship with their biological children. Have the parents grown past the basic narcissistic position which drives procreation? Many parents decide to have children to satisfy their fundamental emotional need to be loved rather than to love, to satisfy societal demands for gender roles, or to attain at least one definition of "success." These parents often cannot comprehend the demands of loving unconditionally and expecting nothing in return until they are faced with a demanding infant 24 hours a day, 7 days a week. Some are not able to cope with these demands successfully. If they have not been up to the task with their biological children, they may not be good candidates for fostering or adoption, which has numerous additional issues beyond the routine demands.

A strong marriage and parenting "partnership" is another desired characteristic. Parents must be able to work through and compromise on differing and varying response styles (which almost always exist), communicating effectively and frequently while supporting each other's position. Playing one parent against the other is a common tactic even with healthy children; however, it takes on new meaning with children coming from abusive backgrounds. They have often survived by paying attention to the emotional nuances of the adults in their world and using those nuances to their advantage. Even when one foster/adoptive parent is able to avoid ego involvement, being calm, rational, and consistent, they will not be successful as a family if the other parent is more reactive and dependent on the children's behavior to drive good parenting. Troubled children are more likely to exploit the parents' differences than to be able to cope with them. Parents with differing parenting styles may also experience problems, which will be discussed further in the section on "Special Issues."

The above attributes and abilities are far beyond what many parents typically achieve. Parenting that attains the level of "good enough" and that may work well for biological children is very different from what is needed to successfully promote and maintain good attachment for children in foster care and adoptive placements. The challenge of finding these seemingly perfect family placements appears insurmountable when one considers that our goal is to have these damaged children enter and remain in one placement until they either return to their biological parents or are placed for adoption. Although this is not always possible in practice, it must be our goal in every case if we are serious about limiting the potential for further attachment breaks.

One technique for finding these special families is to target recruitment efforts at populations which may already have a high concentration of effective parents. This might include many who have a sense of mission due to religious or spiritual beliefs, who feel that they have been fortunate and are seeking to give something back to the larger society. Related fields such as teaching, medicine, and social services may yield parents with a higher level of knowledge about developmental issues and specialized parenting traits. This is not to rule out people in other fields of work, but is simply an effort to target specific groups with apparent high potential. One of our agency's most successful placements has been with a couple who had no biological children; however, both are special education teachers with the requisite training and patience to manage a child with multiple issues. Other possibilities are people who have already successfully raised children and are still young enough to parent again, those who may be involved in kinship care, and those who have a strong and demonstrated commitment to community service.

Although it is certainly possible for newly trained foster or adoptive parents to be successful, attachment-disordered children have been known to "chew up" the new and uninitiated. The behavior of children with attachment problems can be extreme and outside the expectations of an inexperienced and idealistic family with little knowledge of the realities of the child's previous life. These children can and do share some of the behavioral traits seen in other troubled children (tantrums, whining, defiance, aggressiveness, etc.). However, those with the poorest attachments often demonstrate much more severe behaviors, including stealing or hoarding food, self-injury, assault, chronic stealing, pathological lying, fire-setting, cruelty to animals and/or children, and complete lack of guilt or remorse about their behavior (Delaney, 1991). Since every move increases the risk for children with fragile or poor attachment connections, parents who have demonstrated success in working with difficult children (either professionally or in their own families) may provide the best hope of successful placement.

In interviewing or home-studying potential families, it is also helpful to have a checklist of characteristics that lend themselves to positive parenting of these children. Some of the ones which we have found to be most crucial are

- flexibility
- ability to stay calm and matter-of-fact in the face of unreasonable behavior
- well-grounded, not ego-invested in being "right"
- ability to choose appropriate and logical consequences.
- interested in a parenting "challenge" without being invested in "winning"
- ability to give emotionally without expectations of immediate reciprocation
- looks for emotional support from other adults, not the child
- ability to separate the child from the behavior
- a good sense of humor in which the ability to laugh at self is evident; humor is not based in sarcasm, hostility, or put-downs

TRAINING AND SUPPORT OF FOSTER/ADOPTIVE PARENTS

As we have already discussed, given the level of damage to children entering the foster care system, foster and adoptive parents must be adequately equipped to handle the variety of developmental, emotional, and physical needs and deficits of the children in their care. The skills needed to work with attachment-disordered children will work well with any child, but the skills needed to work with nonattachment-disordered children may not be sufficient in reverse. If ill-prepared, parents face a task which is overwhelming and potentially destructive to their own biological family and marital relationship, causing them to abandon fostering or adopting and leaving the system of care even more impoverished. If we have followed our stated principles in recruiting, we will have a pool of parents who are above average in their parenting skills. However, because attachment-disordered children vehemently reject love and nurturing, they pose a challenge which may be beyond the skills of even the most well-intentioned parent. Since many of these children did not receive the structure and nurturance appropriate to their developmental needs or capabilities as infants/toddlers, they now do not feel safe or secure with any caregiver. Foster and adoptive parents must be able to recreate those early bonding experiences and feelings in order to gain trust and give the child a sense of security; however, they are dealing with an older, less cooperative, and less physically manageable child. This requires that parents have a high degree of patience, maturity, and the emotional balance which comes from having resolved any childhood issues of their own (Levy & Orlans, 1998).

Most, if not all states, require training programs for foster and adoptive parents. One example is the Model Approach to Partnership in Parenting (MAPP) curriculum required in our state (Florida). These programs give prospective parents a good overview of basic issues relevant to working with foster children; however, they may not go into sufficient depth or give enough specific techniques to be truly useful in helping parents determine how to handle any given situation. In addition, the reality of a child cursing and spitting in one's face can produce significantly different feelings and reactions than a discussion or role play in class. These classes are a valuable foundation; however, they must be supplemented with more comprehensive training related to child development and attachment issues, such as the following

- the process of child development
- normal developmental milestones
- early bonding, memory, and brain development
- the hallmarks of developmental and emotional delay
- general attachment theory
- attunement behavior
- attunement techniques which will promote attachment
- behavior management techniques developed specifically for use with attachment-disordered children

Comprehensive training must be coupled with therapeutic support and frequent interaction with foster and adoptive families in order to avoid disruptions. Parents need a variety of tools and techniques to help them encourage and support attachment while still managing the negative behaviors that distance and avoid. They also need a support system that attends to and validates both their efforts and their feelings of frustration, pain, and rejection. Children who are poorly attached do not often give much back in the emotional arena and parents can end up feeling seriously depleted. Melinda X. was onc of these "distancing" children:

> Melinda was removed from her biological parents and placed with the S. family when she was 3.5 years old. She was still in diapers and feeding from a bottle. She spoke baby talk and expressed numerous unexplained fears and phobias. What is known about her family of origin is that both parents were substance abusers, and were given to frequent bouts of domestic violence. Melinda was a witness to these episodes and was often left alone while her parents were out looking for drugs.
>
> Mr. and Mrs. S. were confident and experienced parents, having raised six children between them and having also been foster parents for several years at the time Melinda was placed with them. In addition, both had survived significant losses in their own lives, preparing them to cope with the needs and losses of the children placed in their care. Despite this wealth of experience, the demands of Melinda's needy and controlling behavior coupled with her inability to connect emotionally was often a draining experience. She was bright, charming and manipulative with other adults, yet she alternated between clinginess and defiance with her foster parents. She was frequently abusive to younger children in the home. Mr. and Mrs. S. eloquently described the struggle inherent in constantly giving emotionally with little hope of anything in return, of waiting years to see even one genuine expression of connection to them.
>
> When Melinda entered therapy shortly after placement, Mr. and Mrs. S. were included in all phases of treatment planning. They read selected materials on attachment, met regularly with the therapist to discuss problems and progress, and had open access to staff for support and encouragement. This regular interaction was vital in providing additional tools and techniques for managing Melinda's specific behaviors. From their perspective, it was equally as important that we supplied a sounding board for their frustrations and acknowledged the difficulty of their task while encouraging their commitment. Mr. and Mrs. S., who subsequently adopted Melinda, have expressed the opinion that they might never have gone past fostering her if not for the intensive and comprehensive support and reassurance they received in fulfilling their role as Melinda's parents.
>
> After two and a half years, therapist and parents agreed that, although Melinda had made some progress behaviorally, she still did not appear to be making the emotional connections expected by that point. A joint decision was made to seek intensive treatment with a specialized program at the Attachment Center in Evergreen, Colorado. Our therapist accompanied the family and participated in the treatment program. Again the family received as much attention and support as Melinda. While Melinda was placed with a specially trained therapeutic foster family for the two week stay, Mr. and Mrs. S. learned additional parenting techniques and were included in every step of decision-making regarding therapy. Their roles were reinforced, their feelings were acknowledged and validated, and they were encouraged and supported by both therapists and the therapeutic foster parents. This was reinforced after they returned home through regular phone contact with the therapeutic foster parents and the Attachment Center therapist, as well as with Melinda's agency therapist.

Today, 2 years later, Melinda is doing well. Her adoptive mother describes her behavior issues as, for the most part, those of any 9-year-old child. Mrs. S. has described with great joy the first few times when Melinda has expressed real emotion and affection for her. Mrs. S. has also commented that, without the support and perspective provided, she and her husband might have been too overwhelmed to even differentiate Melinda's gradual improvement. Although no one can predict the future with any certainty, Melinda's adoption appears to be secure and her attachment to Mr. and Mrs. S. complete. This is a testament to the S. family's extraordinary commitment and resiliency, and to the fact that they did not feel alone in their struggle to include Melinda in their family.

The many issues that can arise with foster children do not occur conveniently between the typical office hours of 9 A.M. to 5 P.M. Staff assistance must be available at all times so that foster parents are not faced with a sense of aloneness in coping with often difficult situations. Providing this level of support is a significant challenge to both staff and agency, requiring a creative deployment of both people and resources. Specific suggestions for accomplishing these objectives will be discussed in the section on "Structural Issues for the Agency," but support must include at a minimum on-call staff availability, access to respite care, and peer mentorship for foster parents.

An often-overlooked tool that is essential to have and to encourage in foster and adoptive parents is a healthy sense of humor. We know that humor is an important element in handling stress. Parents who learn to see the irony or humor in a situation are better able to keep their perspective and their tempers. In addition, if they begin to understand that they can use that same humor to help them manage the child's behavior, they will have less tendency to get locked into power struggles. Children who have been abused and neglected are often accustomed to meeting their needs through manipulation and/or the provocation of their caretakers' emotions. When faced with calm, firm, and loving humor in response to their behavior, they are often quite startled and uncertain of what to do. A calm and smiling "Nice try, Janie!" in response to a manipulative ploy may stop Janie cold. This type of response also tells the child that the parent is calm, in control, and confident in the role of parent, which is essential information to a child who needs to build trust and attachment with a parent figure.

A cautionary note must be introduced at this point. If the humor used is in any way hostile or sarcastic, it will not have the desired effect. In those forms, it is simply a different manifestation of the parent's anger rather than a healthy expression of humor. It is important that, in our training of foster and adoptive parents, we determine that they understand this distinction and observe that they can use genuine humor successfully with each other before we recommend that they use it with the children in their care. Hostile humor will only convince these children that they were right not to trust adult caregivers.

SPECIAL ISSUES

There are several special issues that do not fit in any of the other categories, but merit consideration here because of the high degree of their impact on the success or failure of foster and adoptive placements. These issues of transition time, conflicting parenting styles, parents' family of origin issues, and family fantasies must be addressed throughout the recruiting, training, and support of foster and adoptive families as they are issues which are some of the most difficult and sensitive to manage.

Transition Time

It is unfortunate that children generally enter the child welfare system abruptly because they are in imminent danger. Perhaps there is little we can do to ease that initial move under such circumstances. However, if we are truly committed to preserving and/or promoting healthy attachment, we must be willing to spend the transition time necessary to help children feel safe and secure prior to any further placement. If children are not able to return home, we need to work with foster and adoptive families to plan transitions that make sense for the child as opposed to the convenience of the child welfare system or the agency. The case of Sammy B. illustrates a successful transition which was unusual in that it was entirely geared to the needs of the child as opposed to those of the system:

> Sammy originally came into the system at 5 years old from a chaotic and unsafe home. His parents did little to ensure his return and, after some considerable time, their parental rights were severed. Sammy was very charming, but could also be loud, excitable, and had frequent tantrums. Up to that point, he had seen very little consistency in parenting, so his behavior could only be described as challenging. He experienced several failed placements before being moved to an agency shelter home. Once there, Sammy settled in with a vengeance, announcing to everyone that he was never leaving. He became very attached to the shelter director, and ran in the opposite direction when he saw anyone who might even remotely look like a prospective family. One potential placement visit ended abruptly when Sammy announced to the family that he would jump off their balcony if they didn't return him to the agency that night. Sammy's message was clear—"work at my pace or not at all."
>
> In the search for the right adoptive match for Sammy, his need for transition time and patience was a paramount consideration. Mr. and Mrs. M., both teachers experienced in working with emotionally handicapped children, were interested in Sammy and were subsequently chosen as his prospective adoptive parents. Each step of the process was designed and planned to gradually build Sammy's relationship and sense of security with them. Both Mr. and Mrs. M. were extraordinarily patient and added their own creative touches to each week's plan in order to bolster its success.
>
> Mr. and Mrs. M. started out as volunteers at the shelter, reading and playing with all of the children, deliberately not seeking Sammy out. Over a number of weeks, as he got to know them and gravitate toward them, they returned his interest.

They invited him to go out for ice cream one Saturday, then brought him directly back to the shelter. Subsequently, they would do some activity with Sammy every Saturday, gradually lengthening the time each week. After some weeks of this, Mr. M. came to pick Sammy up one Saturday by himself and announced that they would have to pick Mrs. M. up at home because she had some work she had to finish. When they got to the house, Mrs. M. was out in the front yard playing with a radio-controlled car. Sammy immediately hopped out to play, then wanted to see their house. After touring, they left and went on to their planned activity; however, Sammy (who loves barbecued ribs) had noted that they had a grill and asked if they could have a picnic one day. They agreed and, over the next few weeks, planned a barbecue. It was in this way that Sammy spent his first afternoon at the M. home.

Simultaneous to these events, Mr. and Mrs. M. had bought clothing and prepared a bedroom especially for Sammy. When he asked about the room, they casually told him it was in case he ever wanted to stay overnight. Sammy spent a number of Saturdays at the M. home without further mention of the room; however, one Saturday he was tired after swimming and asked if he could spend the night. After that point, it gradually became Sammy's idea to be a part of the M. family. His overnights increased to weekends, then finally he decided he was ready to pick up the rest of his belongings from the shelter and stay. Sammy has been with his adoptive family for over two years now and is doing extremely well.

From beginning to end, the whole process of Sammy's move from shelter to adoptive family took 3 to 4 months. He was allowed to feel a measure of control over the direction of his life and was able to explore possibilities at his own pace. All decisions were made with Sammy's needs in mind. This type of thought, planning, and willingness to wait is all too rare in the child welfare system. If we are to have any hope of success in special needs adoptions, we must be more humane in giving children the time and support to become comfortable with changing their world.

Conflicting Parenting Styles

As mentioned earlier, there are specific parenting techniques and styles that are desirable for promoting healthy attachment and managing the behaviors that insecurity and poor attachment generate. Parents who have conflicting parenting styles may unintentionally sabotage their own efforts by confusing the child and providing an opportunity for the child to manipulate or play one parent against the other.

Each parent comes to the family with the parenting ideas experienced in his or her own family of origin. These experiences are deeply ingrained and may be difficult to alter, especially if the parent has any unresolved issues surrounding the way he or she was parented. Often a difference in style is not readily apparent in recruiting or training couples. Some couples may not be aware of any conflict in beliefs or styles because they have not yet parented. Others have parented biological children successfully and therefore do not perceive any potential problem in their differences. However, when faced with parenting a child who is insecurely attached, those differences can become monumental and a source of painful conflict between them. One common example is the family in which one parent is highly structured and the other has a more relaxed attitude toward discipline. A child with attachment disorder does not feel secure with the mixed messages he or she receives in this circumstance. Negative behavior is often the result. Consequently, the structured parent ends up feeling unsupported, the relaxed parent feels badgered, and frustration begins to grow. The resulting tension creates more negative behavior from the child, and a vicious cycle is born.

Agency staff must tackle this subject with families before, during, and after the placement of a child in their home. Working with parents to capitalize on the complementary aspects of their individual styles while minimizing areas of potential disaster is essential and often requires as much attention as any therapeutic work with the child.

Parents' Family of Origin Issues

Although most people decide to foster or adopt children with a genuine desire to help, many also have unresolved conflicts from their own childhood which they may be unconsciously trying to reduce or transform through their relationship with a needy child. Unfortunately, depending on the severity of their issues, the parents' own pathology may instead be triggered or exacerbated by the needs that they are trying to contend with in the child. These parents need resolution of their own childhood pain before they can begin to genuinely meet the needs of a child with similar emotional issues.

Training classes such as the MAPP curriculum previously referenced earlier are a good opportunity to listen for issues of abuse, neglect, or other unresolved conflicts in participants' pasts. These classes are designed to elicit emotional content and many people will detail their own childhood experience as a reason for wanting to foster or adopt, perhaps precisely because they were not helped and they want to make a difference in or change the ending for another child's life. It is important for staff to make note of these revelations for future discussion during home study or other one-on-one settings. Much skill is required to help parents see how their own history may be a hindrance to their effectiveness in parenting a damaged child. They must be encouraged to resolve those early conflicts for themselves and not through the child. Only then will they be able to use their experience as a foundation of understanding and empathy with which they can encourage attachment and manage behavior.

Fantasies

Many potential foster and adoptive families have a conscious or unconscious fantasy of falling in love with and/or saving a child from the horrible life he or she has endured to that point. However, this particular fantasy will be abruptly shattered by a child with attachment disorder. Once the honeymoon is over, parents are suddenly confronted with a child who is not only not grateful to be rescued, but may also be actively hostile and defiant. The "love conquers all" fantasy is a particularly difficult problem for agencies. Despite all of the realistic training and all of the stories we share, an astonishing number of people persist in believing that the child will magically be healed by virtue of inclusion in their family.

Although love may be able to conquer all, it will not do so without a generous helping of reality, patience, skill, training, support, and time. The most treacherous work of foster care and special-needs adoption begins once the placement is made, as we help families grow beyond their fantasies and accept the challenge of giving love without the hidden agenda of those fantasy expectations. It is without question our most difficult task.

STRUCTURAL ISSUES FOR THE AGENCY

The child welfare "system" does not exist apart from its elements: the agency or organization where services to clients actually take place. These may be not-for-profit or voluntary agencies or, more frequently, governmental entities that employ social workers to actually provide direct care to children and families. Agency guidelines and policies concerning staff recruitment, training and support, decision-making, perspective on parents versus professionals, concurrent case planning, foster parent support, funding, postplacement services, and community supports all impact significantly on how casework is conducted and on how services to children and families proceed. In each of these areas, there are opportunities to note which can provide the type of environment for children and families in which healthy attachment is supported.

Staff Recruitment, Training, and Support

Agency staff must be selected who can reasonably be expected to become knowledgeable in child development and attachment issues. This means that recruitment efforts need to focus on locating staff with sufficient education and capacity for and willingness to engage in learning about areas with which they may be unfamiliar, since it is unlikely agencies will succeed in locating enough staff who already have mastery of the subject. Once hired, staff need to receive in-depth orientation and training in various stages and tasks of child development, early relationship issues, and attachment. A thorough command of all of these areas is of critical importance no matter what the level of staff, since each interaction with the agency is an opportunity for enriching attachment experiences for the child.

After the staff has commenced working with families, the agency needs to provide a working environment which prevents the staff burnout that can easily result from the load of supporting the child, the foster family, and the biological family. Staff will benefit from supportive clinical supervision, case collaboration with peers, ongoing training for skill enhancement, the ability to utilize vacation time in a flexible manner, and periodic genuine recognition of their efforts.

As discussed briefly in the section on "Training and Support of Foster/Adoptive Parents," agencies need to be creative with resource utilization in order to provide the required level of responsiveness to foster parents while preventing excessive fatigue and burnout in staff. Some strategies for accomplishing these goals include

- a rotation system that guarantees each staff member will have predictable evenings/weekends free from on-call duties
- "compensatory" time off for staff who have an inordinately long or stressful situation to manage
- regular and effective use of supportive clinical supervision
- clear boundaries and expectations about workload, output, and scope of authority in decision-making
- work units that are structured to allow all staff to be familiar with each case, allowing for easier and more effective substitute coverage
- input into agency policy on working hours and conditions wherever practical
- an ongoing quality improvement process that positions staff in the role of problem identification and solution

Decision-Making and Concurrent Case Planning

An important element of staff support is the manner in which decisions regarding case planning and concurrency are addressed. In spite of every effort to always engage in best practices, the complexities of family situations and the child welfare system often mean that difficult and contentious decisions arise in permanency planning and implementation work. Agency staff function best when they know that their insights and assessments of clients and client needs will be supported as difficulties arise. This may include support not to endorse a particular match, foster parent, or adoptive parent when it is evident this is not in the best interest of the child and to resist pressure from the child welfare system and other outside forces to move things in a manner or time frame that is not in the child's best interest or that may be dictated by political rather than clinical considerations.

Perspective on Biological and Foster Parents

It is imperative that staff have a genuine regard for biological and foster families and resist the urge to "live in the sentimental glamour of saving neglected children from wicked parents" (Holmes, 1993, p. 41). Family-friendly policies include bathroom facilities for visiting parents with appropriate equipment for caring for children, convenient hours for staff services, and the opportunity to responsibly express concerns about the agency and staff without fear of reprisals. Viewing the relationship as a partnership which is centered around the best interests of the child and predicating all communication upon an assumption that each adult plays an important role in assisting the child to achieve optimal development is a helpful perspective and one that must be clearly articulated from agency administration if it is to be pervasive in the organization.

Support Services and Community Supports

Foster parents can be thought of as the last line of defense against the overwhelming wave of children who cannot be cared for in their biological families. It is they who bear the relentless burden of caring for needy and difficult children 24 hours a day, 7 days a week. In order to protect this valuable and indispensable resource, we must help our foster parents develop and provide them with a community of supports. We can not expect them to nurture these children in their families without creating a nurturing environment for the foster families themselves. Some elements of a supportive system for foster parents include

- babysitting services
- parent-to-parent peer networks
- in-home counseling or "wrap-around" services
- easy and ready access to responsive staff
- therapists who are qualified to work with attachment
- additional trained foster parents who can provide short-term respite care for an evening or weekend to give struggling foster or adoptive parents a break
- a mentor program in which experienced foster parents teach, listen to, and encourage newer foster parents
- regular continuing education sessions and social events like potluck dinners that promote the sense of community and partnership between parents and staff

CONCLUSION

Chrissy's entrance into the child welfare system may have been a frightening and bewildering experience; however, there is much that we can do to improve the remainder of her stay and make her subsequent permanency move a healthy one. If we have paid attention to the details of permanency planning, Chrissy will experience (perhaps for the first time) secure attachment and a sense of safety while she is in our care. Secure physical and emotional conditions, coupled with parenting that meets her considerable needs, provide an environment which promotes attachment and the completion of her developmental tasks.

The larger a system becomes, the more daunting the task. There are thousands of Chrissys, Hannahs, Melindas, and Sammys. The temptation is to assume that our task is impossible, and that we can only give each child a little, which is never enough. To do so is to risk not only their future, but the future of our world. This is not a multiple-choice test. There is only one right answer—to act and react in the ways we know will build these children rather than destroy. If not us, then who?

REFERENCES

American Psychiatric Association. (1994). *Diagnostic and statistical manual of mental disorders* (4th ed.). Washington, DC: Author.

Boland, M. G., & Oleske, J. (1995). The health care needs of infants and children: An epidemiological perspective. In N. Boyd-Franklin, G. L. Steiner, & M. G. Boland (Eds.), *Children, families and HIV/AIDS: Psychosocial and therapeutic issues* (pp. 19–29). New York: Guilford Press.

Delaney, R. J. (1991). *Fostering changes.* Fort Collins, CO: Walter J. Corbett Publishing.

Ewing, H. (1993). Women, addiction, and the childbearing family: Social context and recovery support. In R. P. Barth, J. Pietrzak, & M. Ramler (Eds.), *Families living with drugs and HIV* (pp. 18–36). New York: Guilford Press.

Gonzales-Ramos, G. & Goldstein, E. G. (1989). Child maltreatment: An overview. In S. M. Ehrenkrantz, E. G. Goldstein, L. Goodman, & J. Seinfeld (Eds.), *Clinical social work with maltreated children and their families: An introduction to practice* (pp. 3–20). New York: New York University Press.

Holmes, J. (1993). *John Bowlby & attachment theory.* London: Routledge.

Hughes, D. A. (1997). *Facilitating developmental attachment.* Northvale, NJ: Jason Aronson.

Karen, R. (1994). *Becoming attached.* New York: Warner Books.

Levy, T. M., & Orlans, M. (1998). *Attachment, trauma, and healing.* Washington, DC: Child Welfare League of America.

Lindsey, D. (1994). *The welfare of children.* New York: Oxford University Press.

Randolph, E. (1997). *Children who shock and surprise.* Kittredge, CO: RFR Publications.

Strassner, S. L. A. (1989). Intervention with maltreating parents who are drug and alcohol abusers. In S. M. Ehrenkrantz, E. G. Goldstein, L. Goodman & J. Seinfeld (Eds.), *Clinical social work with maltreated children and their families: An introduction to practice* (pp. 149–177). New York: New York University Press.

BIOGRAPHIES

Diane Sabatine Steward

Diane Sabatine Steward, L.C.S.W., is a Licensed Clinical Social Worker in Florida. She is a member of the National Association of Social Workers (NASW) and the Association for Play Therapy (APT). Ms. Steward has been providing psychotherapy treatment for 8 years, working exclusively with abused and/or neglected children, their birth families, foster families, and adoptive families. Currently, Ms. Steward is Director of Program Operations at the Intercoastal Division of Children's Home Society of Florida. She oversees the emergency shelter home, the pregnancy counseling, foster care, and adoption programs and maintains a small clinical caseload. She began her career as a secondary education teacher, followed by 15 years as a trainer and Customer Service Manager for Xerox Corporation, then 5 years as a Child and Family Therapist at Kids In Distress. Ms. Steward has taught seminars on dependency and attachment issues in local college classes and community agencies and, most notably, at a summit meeting of Florida dependency court judges in 1997.

Kathryn Rose O'Day

Kathryn Rose O'Day, L.C.S.W., is a Licensed Clinical Social Worker in Florida. She has nearly 20 years of child and family service experience. Currently, Ms. O'Day is Vice-President for Program Development and Evaluation of Children's Home Society of Florida (CHS). In this capacity, she is responsible for working with staff across the state to develop and measure outcomes for a full continuum of services to children and families. Ms. O'Day began her career as a youth counselor working with abused, neglected, and delinquent children at a local police department and then went on to work in the field of addictions with adults, adolescents, and families and with an international teenage-runaway program. Ms. O'Day was Executive Director of the Intercoastal Division of Children's Home Society of Florida for the 4 years prior to her current position. She has served on the Florida Bar's Adoption Task Force and is an adjunct faculty member of Florida International University's School of Social Work, where she teaches evaluative research in social work and other practice courses.

7

ATTACHMENT THEORY: A MAP FOR COUPLES THERAPY

SUSAN JOHNSON

ANN SIMS

University of Ottawa, Canada

It is well understood by followers of attachment theory and research that attachment theory is a theory of the close affectional bonds which remain with us throughout life (Bowlby, 1969). Further, attachment theory can be viewed as a theory of trauma where the absence of or break in secure connection to others is a source of anxiety and traumatic stress (Atkinson, 1997). Indeed, Bowlby (1988) has written that the primary motivating principle in human beings is the need to seek and maintain contact with others. From this perspective the absence of secure connection to significant others creates considerable distress, rendering persons vulnerable to both physical and emotional problems. This chapter takes the position that attachment theory provides a natural theoretical frame for couples therapy interventions when couples complain of conflict and isolation, often accompanied by depression and despair.

One of the main reasons people seek psychotherapy is distress in intimate relationships (Horowitz, 1979). Since the late 1980s the field of couples therapy has expanded considerably, so that now this modality is used not simply to address marital distress but as an alternative treatment for individual problems known to be associated with interpersonal difficulties. Depression and agoraphobia are two such problems (Jacobson, Holtzworth-

Handbook of Attachment Interventions

Munroe, & Schmaling, 1989). It is also used as part of the multidimensional treatments of difficulties such as eating disorders and post-traumatic stress disorder (Barrett & Schwartz, 1987; S. Johnson & Williams-Keeler, 1998).

Couples therapy has long been criticized for lacking a coherent theoretical base, a clear conceptualization of the workings of adult intimacy which can guide clinical intervention. Of the three perspectives on intimate adult relationships found in the literature, analytic, exchange theory, and attachment theory as it applies in adulthood, attachment theory has been identified as the most promising (Hazan & Shaver, 1987; S. Johnson, 1986; O'Leary & Smith, 1991). A sizable body of research on the relevance of this theory to adult relationships now exists (Shaver & Hazan, 1993). Secure attachment has been found to predict such positive aspects of relationship functioning as greater interdependence, commitment, trust, and satisfaction in couples (e.g., Kirpatrick & Davis, 1994; Simpson, 1990), higher levels of support-seeking and -providing (Simpson, Rholes, & Nelligan, 1992), greater intimacy and less withdrawal and verbal aggression (Senchak & Leonard, 1992), more sensitive and appropriate caregiving behaviors (Kunce & Shaver, 1994), and less jealousy (Hazan & Shaver, 1987).

One of the key developments in the field of couples therapy since the late 1980s has been an empirically guided delineation of the nature of marital distress. Three aspects that have been investigated are the experiencing and communication of emotion between partners, patterns of interaction of distressed couples, and attributions each partner makes concerning the other's behaviors. An important question is whether the separation distress described by Bowlby in his writings on attachment relationships matches the portrait of marital distress emerging from the studies since the late 1980s. Specifically, how do the above components of marital distress, now outlined in the literature, correspond to the thinking in adult attachment?

First, marital distress research emphasizes the role emotion plays in distressed interactions. In fact, facial expression of emotion has been found to be a powerful predictor of future relationship quality and stability (Gottman, 1991). Gottman found that couples were more likely to separate when the wife facially expressed disgust and the husband expressed fear. Interactional patterns of the wife complaining and criticizing with the husband defending and withdrawing occurred at the same time. In this literature emotional engagement is emerging as the greatest predictor of relationship satisfaction and stability. Gottman, Coan, Carrere, and Swanson (1998) have concluded that emotional engagement and responsiveness are crucial characteristics of satisfying close relationships. Attachment theorists have also stressed the importance of emotion,

> Many of the most intense emotions arise during the formation, the maintenance, the disruption and the renewal of attachment relationships. The formation of a bond is described as falling in love, maintaining a bond as loving someone and losing a partner as grieving over someone. Similarly, threat of loss arouses anxiety and actual loss gives rise to sorrow, while each of these situations is likely to arouse anger. The unchallenged maintenance of a bond is perceived as a source of security and the renewal of a bond as a source of joy (Bowlby, 1980, p. 40).

Other attachment writers have identified emotion as the primary factor that initiates and organizes interactions in attachment relationships (e.g., Sroufe, 1979). The need for emotional engagement with an attachment figure, particularly in times of stress, is a key tenet of attachment theory.

Secondly, the empirical work on marital distress highlights the patterned interactions that characterize the drama of distressed adult relationships. These interactions form self-perpetuating loops such as pursue–withdraw, or more commonly, criticize-and-complain responded to with defend-and-distance. The research literature does not offer explanations regarding why such patterns are particularly associated with marital disruption. A theory is needed to give meaning to such negative patterns. Bowlby's emphasis on the relevance of actual interaction patterns between attachment partners to relationship functioning distinguished him from Object Relations theorists and created a natural link between attachment and interpersonal, systemic conceptualizations of relationships. From an attachment perspective such patterns emanate from the separation distress that one partner experiences when perceiving the other as inaccessible or unresponsive. The power of these patterns to create and perpetuate distress arises from the thwarting of a wired-in need for safe connection to significant others. In an ongoing process, these patterns elicit fear and insecurity and often prime negative working models of self and other; models that then become the basis of one's experience of the world. When a distressed couple enters a therapist's office, the therapist observes the sequence of angry protest, clinging and proximity-seeking, depression and despair. If there is no response that reestablishes security, the distressed partner, or partners, may then detach, completing the sequence that Bowlby (1969) described in his original formulation of attachment theory. It is not surprising, from an attachment point of view, that distressed couples speak of their distress in terms of life and death. A wife, for example, may turn to her husband and say, "You let me down. You turned away and let me drown."

Third, the literature on marital distress emphasizes the importance of attributions in the creation and maintenance of marital problems. Partners in distressed relationships interpret each others' responses in ways that perpetuate their distress. They tend to make stable, global, and internal attributions for their behavior (Fincham & Bradbury, 1990). When partners are experiencing relationship distress and one of them is late, for example, the other partner may conclude that this reflects a lack of caring and a desire to hurt. From an attachment perspective, it is the negative working model of the other which colors the perception of the partner's behavior and pulls for specific attributions. Attachment working models then guide future interpretations of behaviors and responses to the attachment figure.

Both attachment theory and recent research on the nature of marital distress portray distressed couples as struggling with powerful negative emotions such as fear and anger. These emotions then prime negative models and interpretations of ambiguous social cues and predispose partners to fight, flight or fear responses that tend to perpetuate distress. The abilities that characterize secure attachment relationships, such as the capacity to meta-monitor conversations, remain flexible, and see the other's perspective (Kobak & Cole, 1991), are precisely the qualities referred to in the marital literature as differentiating distressed from nondistressed parties. Gottman (1994), for example, emphasizes the ability to step back and "unlatch" from negative interactions and create new, more empathic patterns of interaction.

ATTACHMENT THEORY APPLIED TO ADULT DISTRESSED RELATIONSHIPS

The field of marital and family therapy has until recently tended to see the basis of a close relationship as a bargain rather than as a bond (S. Johnson, 1986). This conceptualization has directed the therapeutic focus toward helping partners formulate behavioral exchange contracts or learning specific negotiating and communicating skills. The field has also, as Bowlby suggested, tended to pathologize dependency and to focus on concepts such as enmeshment which sometimes confuse caring with coercion (Green & Werner, 1996) so that issues of emotional disengagement and the deprivation of innate needs are not addressed. From an attachment perspective the underlying cause of marital distress is the lack of accessibility and responsiveness of at least one partner and the problematic ways in which the partners deal with their insecurities (S. Johnson, 1996). The way in which one partner struggles with and defends against insecurities tends to perpetuate the insecurities of the other. From this perspective, the goal in couples interventions is to help the couple not only step aside from escalating cycles of attack and defend, but also to orchestrate the creation of a secure bond. In attachment terms this bond provides a safe haven and secure base for each of the partners.

The approach that has most consistently and systematically used attachment theory as a guide to intervention in repairing adult relationships is Emotionally Focused Therapy (EFT). The first outcome study on EFT appeared in the literature in 1985 (S. Johnson & Greenberg, 1985). At present this approach is one of the best delineated and empirically validated approaches to couples therapy (Baucom, Shoham, Mueser, Daiuto, & Stickle, 1998; S. Johnson, Hunsley, Greenberg, & Schindler, 1999). Before considering the EFT model as an example of how attachment theory may guide the couples therapist, it may be useful to consider the general implications of attachment theory for the couples therapist.

The main implications of attachment theory for the couples therapist are as follows:

1. Needs for contact, comfort, security, and closeness are the defining features of the landscape of adult intimate relationships. These relationships will tend to be "unduly influenced by those occasions when one member of a couple is seriously distressed and the other member either provides psychological proximity or fails to do so" (Simpson & Rholes, 1994, p. 22). Relationships are not defined so much by the equity of giving and getting, as exchange theory suggests, but by moments of emotional responsiveness where attachment needs are either met or frustrated. Distressed couples often come to the marital therapist with tales of specific attachment injuries that have not been healed and are undermining the security of the bond between partners. Even in a generally secure relationship with partners who are comfortable with trust and closeness, such injuries can flood the relationship with negative emotions and constrict interactions. These injuries have then to be encountered and resolved in therapy and trust restored. Negative cycles in general tend to revolve around moments of recurring rejection and/or abandonment that perpetuate attachment insecurities. These moments must be attended to in therapy. If the therapist can foster interactions where one partner is able to express vulnerability and attachment needs and the other can respond, such moments are able to redefine the relationship in a very powerful way.

2. An attachment perspective is a nonpathologizing one. Dependency and attachment needs are not seen as problematic in themselves and are best validated in couples therapy. It is the way partners process and enact such needs in a context of perceived danger that leads to relationship distress (S. Johnson, 1996). Echoes of past attachment relationships that create specific sensitivities in a present relationship are acknowledged in therapy, but are not considered in a deterministic frame. Attachment styles, learned in past relationships, can be modified in new ones.

3. Attachment theory stresses the role of affect in close relationships. From the beginnings of life onward, contact with an attachment figure "tranquillizes the nervous system" (Schore, 1994, p. 244). Emotion primes and directs attachment responses; it is the music of the attachment dance. One must not assume that emotional responses can be left in the background and unattended when working for change in a relationship. Indeed, these

responses have the power to easily undermine other mechanisms of change, such as insight or negotiated behavior change. Emotion, rather, can be a powerful ally in the change process since it directs peoples' attention to their own needs and desires and also, when expressed, communicates to others. When relationships become dangerous, emotional signals become distorted and often make responsiveness on the part of the partner more difficult. Fear, in particular, tends to limit information processing and interactional flexibility. It follows that the provision of a safe base and help in processing emotional responses are key parts of the work of couples therapy. The essence of attachment security is that partners can turn to each other for comfort and help each other with the regulation of emotional distress.

4. This perspective suggests that there is a predictable, finite set of responses that will arise when an innate, survival need is frustrated (Main, Kaplan, & Cassidy, 1985). If people cannot elicit responsiveness by direct emotional signals and protest behavior, as is possible for those with secure relationships, attachment responses will be hyperactivated or minimized, depending on the attachment models of the person (Bartholomew & Horowitz, 1991). Partners who are habitually insecure will have internal working models of self and other and ways of responding that make repair of close relationships more difficult. Partners' working models, abstracted from past experience, will guide their construction of present experience. Those with an anxious, hyperactivated style will tend to aggressively demand reassurance and be vigilant for and reactive to threats to the relationship. Partners with an avoidant style will tend to avoid closeness, ignore their attachment needs, and withdraw precisely when these needs arise in them or in their partner (Simpson et al., 1992). To the extent that these insecure models are closed, easily stimulated, and rigidly applied, fostering positive interactions will be more difficult. The couples therapist will have to find ways to address and deal with these individual differences in attachment style (S. Johnson & Whiffen, in press).

5. The couples therapist must inevitably deal, not only with how the other partner is perceived, but with matters of individual self-definition. Bowlby (1969) stressed that attachment relationships are the arenas where the model of self is originally defined and continually developed and refined. As Guidiano (1991) has suggested, we know ourselves always in relation to others, particularly those who are closest to us. Feminist writers from the Stone Center (Jordan, Kaplan, Miller, Stiver, & Surrey, 1991) have also emphasized that self-definition emerges in close, connected relationships rather than primarily in processes of separation and differentiation. Thus, an anxiously attached partner, who tends to see him- or herself as unworthy of love and caring, will often interact in ways that elicit from a spouse feedback that confirms the negative self-view.

6. Different levels of insecurity will influence the process of change in couples therapy. If a couple has positive expectations and a generally secure

relationship or if a couple has a more distressed relationship that is evoking anxious or avoidant response sets, there is evidence that the attachment-oriented interventions used in EFT are effective (S. Johnson et al., 1999). The most challenging couples to change will be those where one partner has experienced a "violation of human connection" (Herman, 1992) and has been not only deprived but also abused in close relationships. These traumatized partners have a greater need for a safe, close relationship. However, they are also likely to show an entrenched inability to trust those they need the most. They face a paradox; their partner is at once the solution to and the source of danger (Main & Hesse, 1990). Such partners may particularly need couples therapy, since without the safe haven provided by a positive relationship healing from previous abuse will be difficult. In addition, present relationship distress will tend to exacerbate the effects of such abuse and maintain symptoms such as numbing and angry hyperarousal. If safe attachment is the natural antidote to such traumatic experiences (van der Kolk, Macfarlane, & Weiseath, 1996), providing a natural healing arena, couples therapy may be a particularly crucial component of treatment (S. Johnson & Williams-Keeler, 1998). The creation of a healing relationship with the partner may require a specific focus on attachment and may have potential to affect individual symptoms in such clients.

THE EFT MODEL

The process of change in EFT takes place in nine steps. The first four involve assessment and the deescalation of problematic interactional cycles. The middle steps (5 to 7) emphasize the creation of specific change events where interactional positions shift and new bonding events occur. In these steps new cycles are shaped that provide an alternative and anidote to the negative cycle. The last two steps of therapy (8 and 9) address the consolidation of change and the integration of these changes into the everyday life of the couple. These steps, although described in linear form, occur in a spiral fashion, where each step incorporates and adds to the next. In a mildly distressed, securely attached couple, the partners work quickly through the steps at a parallel rate. In more distressed couples, the more passive or withdrawn partner is usually invited to go through the steps slightly ahead of the other. The increased emotional engagement of this partner then helps the other, more critical partner shift into a more trusting stance.

The nine steps of EFT are presented below.

Cycle deescalation

Step 1. Assessment. Creating an alliance and articulating the core conflicts and issues using an attachment perspective.

Step 2. Identifying the problematic cycle that maintains attachment insecurity and marital distress.
Step 3. Accessing the unacknowledged emotions underlying interactional positions.
Step 4. Reframing the problem in terms of the cycle, the underlying emotions, and attachment needs.

Changing interactional positions
Step 5. Promoting identification with disowned attachment needs and aspects of self, and integrating these into relationship patterns.
Step 6. Promoting the other's acceptance of these needs and aspects of self.
Step 7. Facilitating the expression of specific needs and wants and creating emotional engagement.

Key change events, withdrawer reengagement, and blamer softening, evolve here. These events are completed in Step 7. When both partners complete this step a prototypical bonding event occurs, either at home or in the session. In such an event a previously unresponsive partner will reach out and offer comfort and reassurance and the other, more critical, spouse will ask for his/her attachment needs to be met from a position of vulnerability. The relationship is then defined by this interaction as a safe haven.

Consolidation/integration
Step 8. Facilitating the emergence of new solutions to old problematic relationship issues in an atmosphere of collaboration and safety.
Step 9. Consolidating new positions characterized by accessibility and responsiveness and new cycles of attachment behaviors.

In all of these steps, the therapist moves between helping partners crystallize their emotional experience in the present and setting interactional tasks that add new elements to the interactional cycle. The therapist will track, reflect, and expand the inner experience of an individual partner and then use the expression of this experience to create a new dialogue and so reorganize the interactional cycle. The therapist might first help a withdrawn avoidant spouse formulate his sense of paralyzed helplessness that primes his withdrawal, then help his partner to hear his experience, and, finally, move to restructure an interaction around this helplessness. An example could be, "So can you tell her please, I feel like I can never win. I won't grovel. I just want to numb out and run away." The content of this comment is about taking distance. However, the process of confiding represents a move away from avoidant withdrawal and the beginning of active engagement. This partner is then engaged intrapersonally with his own attachment needs and longings and interpersonally with his spouse.

In terms of the EFT change process, the first shift, which occurs at the end of Step 4, constitutes a first-order change (Watzlawick, Weakland, & Fisch, 1974). Reactive emotional responses that heighten each partner's attachment insecurities are still in place but they are less extreme. The partner begins to risk more emotional engagement and to view the cycle or dance, rather than the other spouse, as the enemy. This shift sets the stage for the work of second-order change, reorganizing the interactional dance in the direction of safe attachment. In the middle stage of therapy (Steps 5–7) there are two change events that are crucial turning points in EFT. The first is the withdrawer's reengagement where this partner becomes more active in defining the relationship and more accessible to the other partner. The second change event, a softening, occurs when a previously critical spouse is able to risk expressing attachment needs and fears and to begin to trustingly engage with his or her partner. Research on the process of change has found that this event predicts recovery from marital distress in EFT. If there is only partial engagement in these change events, or if the couple reaches an impasse here, the relationship may still improve but the impact of therapy will be less potent. Transcripts and detailed descriptions of a softening, which often ends in a prototypical bonding event where one partner asks for comfort from a responsive other, can be found in the literature (S. Johnson & Greenberg, 1995; S. Johnson, 1996).

The bonding event that usually occurs in the final stages of a softening initiates a new cycle of confiding, emotional engagement, and responsiveness. This kind of event has the potential, because of its emotional salience in terms of meeting basic attachment needs, to heal past injuries and wounds in the relationship and to redefine the nature of the bond. When, for example, a critical, aggressive partner, who is in the protest and pursue stages of separation distress, is able to share his or her deep fears of abandonment and receive comfort and reassurance from his or her partner, the bond changes. He or she experiences a shift from isolation to connectedness and from frustration to a sense of efficacy in his or her ability to create a new kind of relationship. Once this kind of change event has occurred the couple moves naturally into consolidating this positive cycle. They are then able to solve practical issues which have been fueling the conflict between them.

Who is EFT best suited for? First, EFT is not used for couples where abuse is an ongoing part of the relationship. Abusive partners are referred to group or individual therapy to help them deal with their anger and abusive behavior. They are offered EFT only after this therapy is completed and their partner no longer feel at risk. Second, EFT is used only in an abbreviated form for couples who are separating. Research on success in

EFT (S. Johnson & Talitman, 1997) also allows us to make some specific predictions regarding who will benefit from EFT. As with most psychotherapies, in this study, the alliance with the therapist predicted outcome. However, it was the perceived relevance of the tasks of therapy that was the most important aspect of this alliance. Task relevance was more central to success than a positive bond with the therapist or a sense of shared goals. In general, the couples who do well in EFT are those for whom the focus on attachment needs and the creation of trust and emotional connection make sense. The positive outcome and the large effect sizes associated with EFT (S. Johnson et al., 1999) suggest that this perspective resonates with the couples who come for couples therapy. The collaborative nature of the alliance with the therapist, which provides the couple with a secure base in therapy, generally promotes active engagement in the therapy process. In the S. Johnson and Talitman (1997) study, this alliance and, by implication, this engagement, proved to be a much more potent predictor of success than initial distress level. This is an interesting result in light of the fact that initial distress has been found to account for up to 46% of the outcome variance in other marital therapies (Whisman & Jacobson, 1990).

Johnson and Talitman also determined that EFT worked well with male partners who were described by their partners as inexpressive. This suggests that EFT may provide such partners with a safe base from which to explore and express their experience of the relationship. It was also more effective with older men (over 35), who may be more receptive to a focus on intimacy and attachment. The most powerful predictor of success was, however, a particular kind of trust on the part of the female partner, namely, the female partner's faith that her spouse cared for her in spite of all the difficulties in the relationship. Given that, in western cultures, the female partner tends to monitor the quality of the bond and take responsibility for maintaining closeness, once this partner loses hope in her spouse's caring, a positive redefinition of the marital bond may not be possible. The end point of Bowlby's model of separation distress is detachment, an unwillingness to risk and invest in the relationship. The research on marital distress similarly finds that emotional disengagement is more predictive of marital dissolution than factors such as the ability to resolve disagreements (Gottman, 1994). An unwillingness to emotionally engage in therapy and with the partner may be a key negative prognostic indicator in any therapy that focuses upon the creation of a more secure bond.

EFT INTERVENTIONS

The therapist focuses upon two tasks, the accessing and reformulating of emotional responses and the shaping of new interactions based on these responses. In the first task, the therapist focuses upon the emotion that is most poignant and salient in terms of attachment needs and fears and that plays a central role in patterns of negative interaction. The therapist stays close to the "leading edge" of the client's experience and uses experiential interventions (Perls, 1973; Rogers, 1951), reflection, evocative questions, validation, heightening, and empathic interpretation, to expand that experience. Reactive responses such as anger evolve into more primary emotions such as a sense of grief or fear. In the second task, the therapist tracks and reflects the patterns of interaction, identifying the negative cycle that constrains and narrows the responses of the partners to each other. The therapist uses structural techniques (Minuchin & Fishman, 1981), such as reframing, and choreographs new relationship events. Problems are reframed in terms of cycles and patterns and in terms of attachment needs and fears. So, for instance, the therapist will ask a partner to share specific fears with his/her partner, thus creating a new kind of dialogue that fosters secure attachment. These tasks and interventions are outlined in detail elsewhere together with transcripts of therapy sessions (S. Johnson, 1996; 1998, 1999).

THE CLINICAL EFFICACY OF EFT

To date four randomized clinical trials of EFT have been conducted. In three other studies, subjects acted as their own controls; in one of these the primary focus was upon predictors of success in EFT (S. Johnson & Talitman, 1996). Two studies have also been conducted with couples whose primary focus was not marital distress (one focused on intimacy problems and one on low sexual desire). All EFT studies have included treatment integrity checks and have had very low attrition rates. In a summary article of EFT outcome research, the effect size from the four clinical trials of EFT was calculated at 1.3 (S. Johnson et al., 1999).

This effect size indicates that the average person treated with EFT is better off than 90% of untreated persons. This exceeds published estimates for the effect sizes of couples therapy interventions that do not use an attachment perspective. Follow-up results suggest that treatment effects are stable or improve over time (S. Johnson et al., 1999). In terms of the percentage of couples recovered (not simply improved but scoring in the nondistressed range), the first and the most recent studies of EFT found rates of 70 to 73% recovery from relationship distress, in 8 to 12 sessions (S. Johnson & Greenberg, 1985; S. Johnson & Talitman, 1997). There is also a number of small studies on the process of change in EFT that support the notion that engagement with emotional experience and interactional shifts are the active ingredients of change in this approach (S. Johnson et al., 1999).

HOW WILL THE PROCESS OF THERAPY DIFFER FOR COUPLES WITH DIFFERENT ATTACHMENT STYLES?

Emotionally Focused Therapy is a clinical intervention that addresses separation distress and attachment insecurity in adults. Attachment theory provides a needed theoretical base for marital therapy interventions, interventions designed to create not just more skilled negotiation, but secure attachment. In the EFT approach the first task of the therapist is to create a secure base in the therapy sessions. Research suggests that if partners believe that their spouse genuinely cares for them it is easier for them to engage with the therapist and the therapeutic process (S. Johnson & Talitman, 1997). In our clinical work we have observed that partners with different attachment tendencies demonstrate differing attitudes and behaviors. In the beginning stages of therapy, avoidant partners, for example, are more likely to be skeptical about therapy and wary of the therapist. They often cannot identify feelings or relationship needs, but simply want conflict and distress to end. They prefer to focus on instrumental issues, which they discuss in detached way. We have noted that these avoidant partners usually grasp the interactional cycle from a meta level, yet tend to remain removed from the impact of cycle. Most often they do not understand the effect of distance on their partner and tend to discount it. They often make disparaging remarks about experiences of dependency and vulnerability in others.

In our clinical observations, anxiously attached partners also demonstrate characteristic responses and behaviors at the beginning of therapy. For example, they tend to show more diffuse, absorbing affect than either more secure or avoidantly attached spouses. They are likely to be more reactive and less coherent in their presentation of their relationship and its problems. They interpret a wide range of relationship events negatively, consistent with their anxious attachment. Finally, traumatized partners, who show disorganized attachment, often vacillate between connecting with the therapist and being dismissive and hostile (Alexander, 1997).

Given the differences in the way persons with different attachment proclivities present at the beginning of therapy, it follows that therapist strategies and interventions will vary with attachment style. With avoidant partners, for example, who have difficulty experiencing and identifying feelings, the therapist has to ask emotionally evocative questions, heighten any emotional response, and probe or suggest responses one step beyond the partner's awareness. The therapist must intervene to enhance the engagement of the avoidant spouse in their own experience and in dialogue with their partner. If an avoidant partner says, "I guess I do wall myself off," the therapist will create emotional engagement by repetition and imagery or by asking evocative questions. She/he will then heighten engagement with the partner by asking the avoidant spouse to share the emerging feelings. The attachment history of an individual is used to validate and legitimate present ways of perceiving and responding to the spouse. An avoidant spouse may, for example, tell the therapist that he absolutely refuses to "put all his eggs in one basket." This response is framed as a courageous adaptation to a world where he feels there is no one he can depend on.

Therapy with a couple containing an anxiously attached partner requires interventions which accommodate to this way of being. The tendency of anxious partners to be reactive and interpret relationship events negatively requires the therapist to validate secondary reactive affect and help differentiate and expand this affect until primary attachment emotions can be stated coherently. Thus, an angry wife's blaming statement, "There's something wrong with him. He just can't love anyone," leads to an exploration of her rage, which eventually results in her articulating her underlying desperation and loneliness. She and the therapist then outline together the impact of this desperation and rage on her partner and the negative cycle that ensues. By framing her negative responses as attachment insecurity and deprivation, the therapy helps her partner reappraise her behavior more sympathetically. The anxious partner often experiences the relationship as chaotic and overwhelming emotionally. The therapist helps her articulate the distress and structure the chaos into a coherent attachment story with the interactional cycle as the villain.

In general, no matter what attachment styles the partners bring to therapy, even in the early stages of EFT, engagement in emotional experience brings forth general beliefs about relationships and specific appraisals of the spouse. These appraisals and beliefs are then available for modification. The couple begins to see the attachment drama both as observers and as players who can change the plot as it evolves. However, as therapy continues, it is perhaps easiest to understand the connection between attachment style and the process of repairing a relationship if one views secure attachment as trust in others and a sense of empowerment or confidence in the self (Antonucci, 1994). The therapeutic challenges of, first, expanding constricted interactional cycles and working models and, second, daring to risk engaging emotionally when facing attachment fears are more readily met when couples are confident and trusting. S. Johnson and Talitman (1997) found that a particular form of trust, faith in the other's caring, was the variable most predictive of success in EFT. This kind of trust can be seen as offering an antidote to the attachment fears that partners experience when their relationship becomes distressed. When there is low trust and considerable fear in the relationship, the therapist must work harder to create a safe haven and a secure base in the therapy session and shape the process into small, manageable steps so that the partners have faith in the possibility for change.

CASE EXAMPLE: TWENTY YEARS OF SCREAMING AND SILENCES

Peter and Diane were a professional couple in their mid-40s married for 25 years with children in college. Diane was seeing an individual therapist for her depression and had also requested couples therapy since she saw her depression as intertwined with her relationship with her husband. They described their relationship as having degenerated in the past 5 years into power struggles and week-long silences. This had continued to the point where Diane was questioning her commitment and thinking of leaving. The beginning years of their marriage they remembered as positive but their problems had escalated 10 years ago, when Diane had become engaged in her own career and had made new friends, often going out in the evenings with them. From her point of view Peter had responded to this by becoming extremely angry and coercive and when this did not succeed in changing her behavior he would then "punish" her by refusing to speak to her for days. A recent incident where she had gone for a drink with a tennis instructor had particularly infuriated Peter and had persuaded him that they needed to seek some professional help.

From an attachment perspective, Peter seemed to exhibit the protest and seeking behaviors of separation distress. He described himself as anxiously attached on the Relationship Questionnaire (RQ)(Bartholomew & Horowitz, 1991). His hypervigilance, demands for his wife's attention and anger, sometimes expressed in verbal attacks and sometimes in withdrawal and refusal to respond, fit with the patterns found in anxious attachment. Anxious partners live in "constant fear of losing significant others" (Simpson & Rholes, 1994, p. 187) and tend to focus on their anxiety and anger, amplifying these emotions and alienating their partner in the process. Diane described herself as fearful avoidant and she seemed to be in the depression phase of separation distress, moving toward detachment. Simpson et al. (1992) point out that avoidant partners inhibit emotion and tend to avoid engagement, particularly when they or their spouse experience vulnerability and need. Diane stated that she did not like "mushy stuff" and would cut off discussions in the first few sessions by announcing that she did not want to discuss an emotional topic further. As a child she had a speech disorder and an associated social phobia; as Shaver and Clarke (1994) note, those who adopt a fearful avoidant style of engagement with others tend to lack social self-confidence. Her ability to assert her need for friends and a career had, as she described it, only emerged after a lengthy struggle. Avoidant partners, in our experience, often state that they have decided that they will never again be hurt in a close relationship and consciously organize their responses with this in mind. Diane remarked that after "twenty years of screaming and silences" she had a wall around her and was permanently "on guard" around Peter.

The question of whether this style was a personality predisposition that would have manifested in any relationship or a reflection of her particular relationship with her husband is perhaps a moot point after 20 years of marriage. Whatever predisposition she may have had to adopt this style had certainly been confirmed in an ongoing manner by her interactions with Peter. Neither partner had been traumatized in close relationships in childhood, but Peter spoke bitterly of isolation and disengagement as the norm in his family. Diane was the only person he had ever felt an intimate connection with. The couple mentioned that Peter's anger had spilled over into one violent incident a few years ago but the couple minimized the significance of this and, as Peter had promised, it had not occurred again. Diane specifically stated that she was not afraid of her partner.

Considering the variables particularly associated with attachment style (S. Johnson and Whiffen, in press), namely affect regulation, information processing, and communication patterns, Peter was hypervigilant and reactive in the first sessions and Diane remained composed and logical, but would occasionally weep silently and turn away, refusing to speak. Diane often asked the therapist for solutions and answers while Peter commented that she was after "formulas" and responded with anger. Diane would then label Peter as having a personality problem and his demands as "endless and insatiable." In terms of attributions, the couple labeled each other and saw the other as trying to dominate and control. Kobak and Cole (1991) suggest that attachment insecurity manifests as a closed diversionary or closed hypervigilant information processing style. In this couple, Diane would interrupt emotional dialogue with meta-comments and demands for solutions and dates for problem resolution from the therapist, while Peter would punctuate intense arguments for his point of view with charged reactions, such as, "If I have to ask for attention you can keep it. Stuff it, I don't need it."

In terms of communication patterns, neither could take the other's perspective or sustain emotional engagement. The cycle was one of pursue and coerce with periods of hostile disengagement on Peter's part and distant defensiveness on Diane's part. These interactions confirmed the others attachment fears and models of self, other, and the relationship and prevented the updating or elaboration of these models. Diane, for example, saw Peter's hostile disengagement as a particularly potent weapon that reduced her to feeling like she was "nothing" and that the relationship was "hopeless" and dangerous. Peter identified his anger in the first sessions as arising from the sense that he was "number 49 on a priority list of 12" with Diane. He felt alone and unimportant. Interchanges such as Peter stating "You are never home" and Diane replying, "You want to control every minute of my life. I want my freedom" fit the patterns Gottman identifies as the destructive four horsemen of the Apocalypse for marriage (1991).

The therapist identified the couples interactional cycle (Step 2 of EFT) and framed it as an attachment drama which left them both feeling desperate and hopeless and unable to get the comfort and closeness they needed. Peter was framed as being highly sensitive to Diane's distance and attempting to hold on to her while Diane felt constricted and resentful and withdrew more. Other approaches to couples therapy might have focused on power and negotiation skills and making contracts regarding the amount of time Diane spent with her friends or insight into family of origin issues. The EFT therapist highlights attachment issues and, from the beginning, helps the couple articulate attachment needs and fears that organize their interactional cycles. The dialogue expanded in the secure base provided by the therapist, who validated both partners perceptions and responses, integrating them together with the emotions implicit in the interaction into a coherent account of the relationship dance. Peter, starting from the position of, "this is against all my instincts. I refuse to expose myself. I can leave if I have to," began to talk about how he felt he was losing Diane and this was like being "a deer caught in a car's headlights." He acknowledged that he was very sensitive to Diane's distance and that he would try to control her. When this did not work he said he would attempt to "cut my losses" and cut off his own need for her by numbing out and withdrawing. Diane saw his rages as abusive and his silence as "cruel" and designed to get her to "grovel" in order to appease him.

As well as identifying the cycle and underlying feelings (Step 3 of EFT), models of self and other were clarified and updated. Peter admitted that he was often preoccupied with his job and stated that, in fact, he did not need Diane to be home all the time waiting for him. He did need to feel that he came first with her and was considered in her plans. He spoke with pain of how he felt "merely tolerated" by Diane and labeled "defective." Diane was amazed by his expressed need for reassurance, seeing him as generally "indestructible." Diane explained her experience in terms typical of more withdrawn partners, that is, as being dominated by a sense of, "I try to be with him but nothing is ever enough so I give up and go away." Peter began to hear how his protests put her "in a vice" and drove her away from him and to see her as unhappy and desperate rather than uncaring. The first four steps of EFT were completed and resulted in the deescalation of the negative cycle of interactions and a concomitant lessening of negative emotional responses. The couple began to listen to each other more, to spend time together, and to make love occasionally. They were able to reassure each other a little more and touch each others' sore places a little less; however, the pattern of the dance was still the same.

In the middle phase of EFT, the goal is to create emotional reengagement and shifts in interactional positions; that is, to support Diane, the more withdrawn avoidant partner, to reengage and to encourage Peter, the more anxiously attached partner, to become less coercive and risk trusting

Diane with his vulnerability. He could then ask for his needs to be met in a manner that would elicit compassion and acceptance from his wife.

As the therapist tracked and highlighted Diane's emotional reactions, she began to talk of how "intimidating" she found her husband, both in terms of his anger and his ability to "frost her out." With the therapists help, Peter was able to risk telling her, "It's too hard to show you I'm hurting. You won't respond. You won't be attentive, so I shut down and numb out. If I say, "here's my heart," I'll find a dagger in it." Diane responded to Peter's disclosure by joking and making fun of his comment, suggesting that he wanted "an attentive little wife." Peter completely withdrew. The therapist explored Diane's response with her and Diane elaborated that she did not see Peter as taking a risk. She viewed him as the "iron man" and as dangerous. Diane then began to explore her fear. First, she spoke of his silences where she felt that "he hates me and I am in hell; he flips, he's suddenly a raging bull or he's indifferent and I feel like nothing." Diane articulated her pain at being shut out by Peter and how difficult it was for her to trust him, even if he seemed to be reaching out to her. She then became very detached and suggested that the relationship had improved more than she had expected and that now was the time to stop therapy. The therapist redirected the process back to Diane's fears and asked,

Therapist: What happens Diane when you start to talk to him about your fear and risk showing that fear to him? (She puts her hands over her throat.)
Diane: It's too dangerous. Even when he reaches for me or holds me, I think, when will he turn and explode or shut me out?
Therapist: You have your hands on your throat, Diane, and you seem very shaky, maybe afraid? (Therapist reflects Diane's nonverbal response and heightens the fear implicit in Diane's response.)
Diane: (She weeps and speaks in a small, low voice.) He tried to choke me—I couldn't breathe. (She is overtaken by weeping.) It was only once, years ago now, but I thought—I couldn't breathe.

The therapist realizes that Diane is describing a specific incident, defined in EFT terms, as an attachment injury. This is an incident that generally defines Peter as dangerous and the relationship as insecure and is used in an ongoing way as a touchstone by Diane to determine the level of risk in the relationship and to prime her self-protective strategies.

Therapist: Peter tried to choke you? (Diane nods.)
Peter: (In a very quiet voice) It was terrible. I lost it. I told her afterwards I'd never touch her like that again and I haven't. But we've never discussed it, after that night (Diane weeps).
Therapist: (To Diane) You thought you were going to die?
Diane: Yes. Some part of me was just crushed, crushed that Peter would. . .that Peter could. . . It comes up as an image for me. . . I feel trapped. . . and I say to myself, 'I'll never let you hurt me like that again, never again.'
Therapist: Can you tell him, 'I won't be trapped, confined, I'll protect myself

because I'm afraid of you.' (Peter looks shocked and pale.) (Therapist explicitly reflects and heightens Diane's position and presents a task that shapes a new kind of dialogue.)

Diane: (She brings her head up and looks at the therapist) Yes, I thought I was going to die. (She turns to Peter) How could you do that if you loved me?

Therapist: He suddenly became terrifying for you. (She nods emphatically.) Peter, what happens for you when Diane speaks like this? (The therapist intends to support Peter to stay emotionally engaged.)

Peter: (He looks at his hand speaks very quietly.) I feel terrible. I guess I've tried to wipe that incident out. She's right, my anger was wild, out of line. It scared me too (he puts his hand over his eyes). (This is a very different Peter than the cool, defended Peter of the first sessions.)

Diane: And you still coerce me, threaten me—now you shut me out, deliberately, you leave me all alone. I'm crushed by that. So when you start to risk a little, I... (she leans back in her chair).

Therapist: When he risks, warnings go off in you and you say to yourself, 'Oh no, I'm not going to trust and come close. I'll not be trapped or left out in the cold again.' Is that it? (Therapist crystallizes Diane's response when closeness looms.)

Diane: Right. I can't bear it. And he knows, he knows how it hurts me and he does it anyway.

Peter: (Peter's withdrawal confirms the message of the attachment injury incident.) (In a quiet, intense voice) Diane, I cut off and go away to stop myself, to put a lid on my anger so it doesn't blow up. I don't know what else to do. I never knew it affected you so much. You always seem so cool.

Therapist: Peter, you're saying that you withdraw to protect Diane from your anger (he nods) and because you feel overwhelmed, yes? (he nods) Kind of helpless? (he nods vigorously). But Diane, you feel "hated," like he is using a weapon against you, shutting you out? And fear comes up for you, you sense that you might be trapped, like you were in that hotel room, or shut out and left all alone? Yes? (she shuts her eyes and weeps and nods her head). (The therapist reflects a key part of the couples cycle on the level of underlying, attachment-salient emotions).

Diane: Yes, Yes. Then I get all confused and sometimes I think I should get out, give up.

Therapist: There's no safety here—part of you wants to run. Can you tell him that? (The therapist reflects her ambivalence, her need to protect herself and structures the dialogue.)

Diane: (Turns to Peter, leans forward, and speaks slowly) I put up a wall. I can't be hurt like that again. I can't live with all this fear, all this fear that you'll get that mad again or keep shutting me out, I can't.

Therapist: Ah-ha. What's happening for you, Peter? You look very sad right now.

Peter: I feel sad. (He turns to Diane) You have a right. I get desperate. I need reassurance and I push and push. I know. You look so strong to me. I feel so desperate. I never knew my silences hurt you that much.

Therapist: (Very softly) It's sad?

Peter: Yeah. I don't want her to be afraid. (Therapist motions him to tell his wife.) I've hurt you. I'm so sorry. I'm sorry. I'm sorry. I feel your reserve, your distance and I loose perspective, I'm sorry (he reaches for her hand).

After this session Diane becomes more open and engaged in the relationship and asserts her need to see her friends but in a way that acknowledges Peter's place in her life. Peter then finds he feels less like the "default option" and begins to handle his insecurities differently. He experiences less anger and feels less helpless. He is also able to be responsive and reassuring when Diane says that she needs time and a sense of safety before she can open up to him.

Peter then moves into Steps 5 and 7 of EFT and begins to struggle with his own blaming stance and access his attachment needs and fears.

Peter:	(To Diane) I feel you're holding back. I understand it but it's hard to tolerate. I'm on edge. It's like a test and I don't do auditions (he sets his mouth in a thin line of defiance).
Therapist:	What's happening to you as you say that Peter? It's hard for you to deal with distance, her refusal to stay in the old patterns of your relationship? (Therapist focuses on the attachment relevant emotion, that is, his difficulty with her distance, staying close to his experience in the hope that he will go into it further rather than into his defiance).
Peter:	I feel exposed. It goes against every instinct I have. I feel foolish. Like I'm asking for it. My family tested me all the time and then never turned up when I made it or when I needed help. So, I guess, I push her, I criticize her.
Therapist:	It goes against your instincts to show your soft places; your instinct is to protect yourself, to be in control, not to expose yourself. (Therapist validates his self protectiveness).
Peter:	Right. And I feel judged, like she's telling me I'm defective (his eyes fill with tears).
Therapist:	That's very hard for you—to feel exposed, tested, not to feel powerful. But you are still here, struggling to hold on to Diane (Peter closes his eyes and nods). You get pulled into raging or shutting down. It's hard to feel that fear, that you are vulnerable, that you need her so much. (Therapist reflects his need to protect himself but adds a focus on his attachment fears.)
Peter:	(To the therapist) Yes, I guess, maybe—I need reassurance so desperately (therapist nods). In those silences, I want to come in from the cold, but I get stuck there.
Therapist:	You get scared and frozen, huh? It's so hard for you to show Diane that you need her when you are afraid (the therapist heightens his image and his attachment needs and vulnerabilities)....
Peter:	(Cuts in) I can't. I know she has a right to be reserved, to have her doubts. I'm still here, I'm still trying.
Therapist:	You love her enough that you are hanging in and fighting for this relationship, fighting your own fear and rage (nods), fighting the voice that says 'never expose your vulnerability, your need, never ask', huh? That takes courage. (The therapist validates his struggle with his fear and his attachment to his wife.)
Diane:	(Soft voice, looking at him) I see him trying. I want his arms round me. I know it's hard to risk.
Therapist:	(To Peter) Can you tell Diane right now, Peter, can you tell her how much you need her reassurance and how scary it is? (Therapist sets an interpersonal task that involves confiding attachment needs and fears.)

Peter: I can't say 'scared,' I can say worried. (The therapist encourages him to begin with "worried.")

After this session Peter was able to express the insecurities that arose in him when Diane announced that she was going to be with her friends and she was able to respond assertively but reassuringly, taking his feelings into account. The couple were able to initiate bonding events characterized by mutual confiding and comforting contact. Steps 8 and 9 of EFT then unfolded and the couple were able, in an atmosphere of safety, to negotiate around time apart and other pragmatic issues that had triggered their negative cycles in the past. They were able to clarify the responses that defined the relationship as a secure base and a safe haven for them and step aside from the self-protective behaviors that maintained each person's attachment insecurity.

This couples change process illustrates the EFT approach to working with intimate bonds. The process focuses upon the interactional patterns that continually confirm attachment fears and evoke styles or ways of dealing with those fears. It also focuses upon attachment needs and the adaptiveness of these needs. Peter is not framed as overdependent or as unskilled. The couple have become caught in a cycle, a form of engagement, that heightens his needs for reassurance and also frustrates them. The couple move from protecting themselves in a manner that elicits the other's attachment fears to helping each other with these fears and vulnerabilities. The possibility of losing those we love and depend on "terrifies us in a primordial way" (S. Johnson, 1997, p. 39) and arouses fight, flight, and freeze responses that constrict the relationship dance. The creation of comfort and responsiveness, however, expands not only the interaction patterns of a relationship but how partners are defined within that relationship. The focus on emotional responses accesses attachment needs and models and allows these models of self and other to be articulated, expanded, and updated. Once the relationship becomes a secure base each partner's ability to see the other's perspective increases and pragmatic issues and differences can be negotiated.

Attachment theory offers the couples therapist more than a map of the defining features of adult love relationships. It offers the therapist a route into the inherent longings and fears that are the music of the dance between intimate partners. The therapist can then use this music to create a dance that promotes self-efficacy, well-being, and resilience.

REFERENCES

Alexander, P. (1997). Application of attachment theory to the study of sexual abuse. *Journal of Consulting and Clinical Psychology, 60,* 185–195.

Antonucci, T. C. (1994). Attachment in adulthood and aging. In M. B. Sperling & W. H. Berman (Eds.), *Attachment in adults* (pp. 257–272). New York: Guilford Press.

Atkinson, L. (1997). Attachment and psychopathology: From laboratory to clinic. In L. Atkinson & K. Zucker (Eds.), *Attachment and psychopathology,* (pp. 3–16). New York: Guilford Press.

Barrett, M. J., & Schwartz, R. (1987). Couple therapy for bulimia. In J. Hansen & J. Harkaway (Eds.), *Eating disorders* (pp. 25–39). Rockville, MD: Aspen Publications.

Bartholomew, K., & Horowitz, L. (1991). Attachment styles among young adults. *Journal of Personality and Social Psychology, 61,* 226–244.

Baucom, D., Shoham, V., Mueser, K., Daiuto, A., & Stickle, T. (1998). Empirically supported couples and family interventions for marital distress and adult mental health problems. *Journal of Consulting and Clinical Psychology, 66,* 53–88.

Bowlby, J. (1969). *Attachment and loss: Vol. 1. Attachment.* New York: Basic Books.

Bowlby, J. (1980). *Attachment and loss: Vol. 3. Loss.* New York: Basic Books.

Bowlby, J. (1988). *A secure base.* New York: Basic Books.

Fincham, F. D., & Bradbury, T. N. (1990). Cognition in marriage: A program of research on attributions. In D. Perlman & W. Jones (Eds.), *Advances in personal relationships.* Bridgeport, CT: JAI Press.

Gottman, J. (1991). Predicting the longitudinal course of marriages. *Journal of Marital and Family Therapy, 17,* 3–7.

Gottman, J. (1994). *What predicts divorce*? Hillsdale NJ: Erlbaum.

Gottman, J., Coan, J., Carrere, S., & Swanson, C. (1998). Predicting marital happiness and stability from newlywed interactions. *Journal of Marriage and the Family, 60,* 5–22.

Green, R. & Werner, P. (1996). Intrusiveness and closeness-caregiving: Rethinking the concept of family enmeshment. *Family Process, 35,* 115–153.

Guidiano, V. F. (1991). *The self in focus.* New York: Guilford Press.

Hazan, C., & Shaver, P. (1987). Conceptualizing romantic love as an attachment process. *Journal of Personality and Social Psychology, 52,* 511–524.

Herman, J. (1992). *Trauma and recovery.* New York: Basic Books.

Horowitz, H. (1979). The cognitive structure of interpersonal problems treated in psychotherapy. *Journal of Consulting and Clinical Psychology, 47,* 5–15.

Jacobson, J., Holtzworth-Munroe, A., & Schmaling, K. (1989). Marital therapy and spouse involvement in the treatment of depression, agoraphobia and alcoholism. *Journal of Consulting and Clinical Psychology, 57,* 5–10.

Johnson, S. (1986). Bonds and bargains: Relationship paradigms and their significance for marital therapy. *Journal of Marital and Family Therapy, 12,* 259–267.

Johnson, S. M. (1996). *The practice of emotionally focused marital therapy: Creating connection.* New York: Brunner/Mazel (now Taylor & Francis).

Johnson, S. M. (1997 September/October). The biology of love. *Family Therapy Networker,* pp. 37–41.

Johnson, S. M. (1998). Emotionally focused couple therapy. In F. Dattilio (Ed.), *Case studies in couples and family therapy* (pp. 450–479). New York: Guilford Press.

Johnson, S. M. (1999). Emotionally focused couples therapy: Straight to the heart. In J. Donovan (Ed.), pp. 13–42. *Short term couple therapy.* New York: Guilford Press.

Johnson, S. M., & Greenberg, L. S. (1985). Differential effects of experiential and problem solving interventions in resolving marital conflict. *Journal of Consulting and Clinical Psychology, 53,* 175–184.

Johnson, S. M., & Greenberg, L. (1995). The emotionally focused approach to problems in adult attachment. In N. S. Jacobson & A. S. Gurman, (Eds.), *The clinical handbook of marital therapy* (2nd ed. pp. 121–141). New York: Guilford Press.

Johnson, S., Hunsley, J., Greenberg, L., & Schlindler, D. (1999). Emotionally focused couples therapy: Status and challenges. *Clinical Psychology: Science and Practice, 6,* 67–79.

Johnson, S. M., & Talitman, E. (1997). Predictors of outcome in emotionally focused marital therapy. *Journal of Marital and Family Therapy, 23,* 135–152.

Johnson, S.M., & Whiffen, V. (in press). Made to measure: Adapting emotionally focused couples therapy to partner's attachment styles. *Clinical Psychology: Science and Practice.*

Johnson, S. M., & Williams-Keeler, L. (1998). Creating healing relationships for couples dealing with trauma: The use of emotionally focused marital therapy. *Journal of Marital and Family Therapy, 24,* 25–40.

Jordan, J., Kaplan, A., Miller, J., Stiver, I., & Surrey, J. (1991). *Women's growth in connection: Writings from the Stone Centre.* New York: Guilford Press.

Kirpatrick, L. E., & Davis, K. E. (1994). Attachment style, gender, and relationship stability: A longitudinal analysis. *Journal of Personality and Social Psychology, 66,* 502–512.

Kobak, R., & Cole, H. (1991). Attachment and meta-monitoring: Implications for adolescent autonomy and psychopathology. In D. Cicchetti & S. Toth (Eds.), *Disorders and dysfunctions of the self* (pp. 267–297). Rochester, NY: University of Rochester Press.

Kunce, L. J., & Shaver, P. (1994). An attachment theoretical approach to care-giving in romantic relationships. In K. Bartholomew & D. Perlman (Eds.), *Advances in personal relationships.* (Vol. 5, pp. 205–237). London: Jessica Kingsley.

Main, M., & Hesse, E. (1990). Parent's unresolved traumatic experiences are related to infant disorganized attachment status. In M. Greenberg & D. Cicchetti (Eds.), *Attachment in the preschool years.* pp. 152–174. Chicago, University of Chicago Press.

Main, M., Kaplan, N., & Cassidy, J. (1985). Security in infancy, childhood and adulthood: A move to the level of representation. *Monographs of the Society for Research in Child Development, 50*(1-2, Serial No. 209), 66–104.

Minuchin, S., & Fishman, H. C. (1981). *Family therapy techniques.* Cambridge, MA: Harvard University Press.

O'Leary, K. D., & Smith, D. A. (1991). Marital interactions. *Annual Review of Psychology, 42,* 191–212.

Perls, F. (1973). *The gestalt approach and eyewitness to therapy.* San Francisco: Science & Behavior Books.

Rogers, C. (1951). *Client-centered therapy.* Boston: Houghton Mifflin.

Schore, A. N. (1994). *Affect regulation and the organization of self.* Hillsdale, NJ: Erlbaum.

Senchak, M., & Leonard, K. E. (1992). Attachment styles and marital adjustment among newly wed couples. *Journal of Social and Personal Relationships, 9,* 51–64.

Shaver, P., & Clarke, C. (1994). The psycho dynamics of adult romantic attachment. In J. Masling and R. Borstein (Eds.), *Empirical perspectives on object relations theory,* (pp. 105–156). Washington, DC: APA Press.

Shaver, P., & Hazan, C. (1993). Adult romantic attachment: Theory and evidence. In D. Perlman & W. Jones (Eds.), *Advances in personal relationships* (Vol. 4, pp. 29–70). London & PA: Jessica Kingsley.

Simpson, J., (1990). The influence of attachment style on romantic relationships. *Journal of Personality and Social Psychology, 59,* 971–980.

Simpson, J., & Rholes, W. (1994). Stress and secure base relationships in adulthood. In K. Bartholomew & D. Perlman (Eds.), *Attachment processes in adulthood,* (pp. 181–204). London, & PA: Jessica Kingsley.

Simpson, J., Rholes, W., & Nelligan, J. (1992). Support seeking and support giving within couples in an anxiety provoking situation: The role of attachment styles. *Journal of Personality and Social Psychology, 62,* 434–446.

Sroufe, A. (1979). Socioemotional development. In J. D. Osofsky (Ed.), *Handbook of infant development* (pp. 462–516). New York: Wiley.

van der Kolk, B. A., Macfarlane, A., & Weisaeth, L. (1996). *Traumatic stress.* New York: Guilford Press.

Watzlawick, P., Weakland, J., & Fisch, R. (1974). *Change: Principles of problem formation and problem resolution.* New York: Norton.

Whisman, M. A., & Jacobson, N. S. (1990). Power, marital satisfaction and response to marital therapy. *Journal of Family Psychology, 4,* 202–212.

BIOGRAPHIES

Ann Sims

Ann Sims is a therapist and doctoral candidate in clinical psychology at the University of Ottawa. Her area of research is the interaction between attachment dimensions and relationship changes through the process of couples therapy.

Sue Johnson

Dr. Sue Johnson is Professor of Psychology and Psychiatry at Ottawa University and Director of the Marital and Family Clinic at the Ottawa Civic Hospital. She received her doctorate in Counselling Psychology from the University of British Columbia in 1984. She is a registered psychologist in the province of Ontario, Canada and a member of the editorial board of the Journal of Marital and Family Therapy. She is one of the originators of Emotionally Focused Couples Therapy (EFT), now one of the best validated couples interventions in North America. In the past 10 years she has authored two books on EFT, *Emotionally Focused Therapy for Couples* (1988, Guilford Press) with Leslie Greenberg and, most recently, *The Practice of Emotionally Focused Marital Therapy: Creating Connection* (1996, Brunner/Mazel). She was also senior editor of *The Heart of the Matter: Perspectives on Emotion in Marital Therapy* (1994, Brunner/Mazel). She has also authored numerous articles and research studies on couples therapy. Sue Johnson is an Approved Supervisor for the American Association of Marriage and Family Therapy and is internationally known for her workshops and presentations on practice, theory, and research in marital therapy. She maintains a private practice and lives in Ottawa, the capital of Canada, with her husband and two children.

8

CORRELATION BETWEEN CHILDHOOD BIPOLAR I DISORDER AND REACTIVE ATTACHMENT DISORDER, DISINHIBITED TYPE

JOHN F. ALSTON

Child, Adolescent, Family, and Adult Psychiatrist
Private Practice
Evergreen, Colorado

ACCURATE RECOGNITION OF MOOD DISORDERS CREATES EFFECTIVE MEDICAL TREATMENT PLANS IN MALTREATED CHILDREN

Case History

Jim (not his real name) was a 9-year-old boy referred for a psychiatric evaluation by his psychotherapist of 1-1/2 years. At the time of this evaluation, Jim had lived with his adoptive parents for a 2-year period. Prior to his adoption, he had been in a foster/adoptive situation with these same parents for 1-1/2 years. He had been in another foster home since relinquishment from his birth mother when he was 2 years of age.

Jim's symptoms were of both a severe and long-lasting nature, predating even the initial relinquishment when he was 2 years of age. These symptoms included "acting out." He was described as stealing, lying, easily agitated, disruptive, intrusive, and verbally and physically aggressive. He was bossy and pushy with his friends. He was further described as self-centered and selfish. Adoptive parents described "up and down cycles" of roughly 3-month intervals from at least the last 3-year period. In his "up cycle," Jim was described as a relatively easy-to-please goal-directed child who atempted to do his best to fit into his new family. On his "down cycles," all of the above-mentioned symptoms were prominent. Additionally, he had a consistently difficult time falling asleep. School and friendships were very difficult because of ongoing oppositional and defiant tendencies. He was described as having no close friends.

Historically, Jim had been relinquished, along with his brother, 3 years older, and sister, 1 year younger, from his birth mother following a Dependency and Neglect Petition when Jim was 2 years old. He was observed on many occasions crawling in the driveway as an infant. The three children had a game of hiding behind parked cars and running out in traffic to scare motorists. He was observed to be completely out of parental control by neighbors, social services, and police on many occasions.

The birth mother, Betty (not her real name), was one of five children born to an alcoholic mother who abandoned the family when Betty was in junior high school. Betty's father had been physically abusive. He had been sexually abusive to Betty's older sister but Betty denied any incest. Betty had run away from foster care at age 14 and met Vince (not his real name). She became pregnant with Jim's older brother. Betty stayed with Vince despite a chaotic marital situation. Vince was physically, verbally, and sexually abusive to Betty. He had "sold her" (sexually) on several occasions in order to obtain drugs and alcohol. Vince had an extensive criminal history beginning when he was a juvenile. Vince came from an extended history of drug and alcohol abuse, sexual abuse, and domestic violence in his own family of origin.

At the time of the initial psychiatric evaluation, Jim was described as having responded to the psychotherapist, who specialized in bonding and attachment disorders. He was described as having been "bonded" with a good "conscience." He was described as having clear periods of being loving and affectionate but at other times demonstrated high degrees of oversensitivity, irritability, and oppositional and defiant traits.

For the most recent 3-year period of time, Jim had been diagnosed and treated for Attention Deficit Disorder by his family doctor and had been taking methylphenidate (Ritalin). At the time of the initial psychiatric evaluation, he was taking this medication in high doses with minimal benefits.

Given the above personal and family history, symptoms, and mental status derived from a personal interview with Jim, a working diagnosis was made of Bipolar Disorder, Moderate, Most Recent Episode Hypomanic (American Psychiatric Association, 1994). A diagnosis was also made of Reactive Attachment Disorder, Disinhibited Type, In Partial Remission (American Psychiatric Association, 1994). A recommendation was made for a relatively rapid decrease to discontinuation of Jim's Ritalin and an initiation of a clinical trial of lithium (Lithobid) in 300-mg tablets in gradual increasing doses up to one tablet three times daily.

On follow-up 6 weeks later, Jim was described by his adoptive parents as "a lot better" and a "completely different kid." Spontaneous words that were used to describe Jim included "calm, polite, settled at night, helpful, loving, considerate, and sharing." He was described as emotionally upbeat and excitable in appropriate ways.

Jim continued doing well for several weeks, but the parents and school noted continued difficulties with motivation, attention, forgetfulness, and task completion. As a result of these concerns, a clinical trial was begun on bupropion (Wellbutrin). This medication was prescribed in modest doses. On follow-up, Jim's concentration, attention, motivation, memory, and task completion were all described as substantially improved. He continued to be described as "even keeled" with no temper outbursts and a greatly improved willingness to please.

Jim continued to be followed-up regularly. He continued to do very well for a 2-year period of time. He excelled in several sports but his classroom esteem, despite overall improvements, was consistently low. By the age of 12, Jim's symptoms had taken a turn for the worse. He was described as increasingly oppositional, "reckless and verbally and physically aggressive to younger children." His previous positive competitiveness in sports had become excessive and inappropriate. As a result of these, and other symptoms, a clinical trial was initiated of risperidone (Risperdal) in 1-mg tablets in increasing doses up to 1 tablet twice daily. On subsequent follow-ups, Jim has been described as doing "the best ever." He was displaying no temper outbursts or other aggressive tendencies. He was described as dealing with "all of his issues in a verbally appropriate manner." He was described as having positive self-esteem and motivation and exhibiting appropriate emotional and behavioral self-control. He has had no side-effects from this medication regime. His prognosis has become excellent with supportive psychotherapy and medication management.

Reconceptualization of Mood Disorders Affecting Maltreated Children

This case history is representative of a large number of cases in which recognition, diagnosis, and treatment of childhood mood disorders have greatly enhanced the quality of life of attachment-disordered children and their families. Medical treatment plans for maltreated children are capable of either promoting or inhibiting a child's well-being. Medications, with the help of other effective treatment techniques, are capable of enormously enhancing the child's abilities to demonstrate attachment behavior, or conversely, contribute to a child's lack of sufficient progress in developing attachment behavior. Many children have made dramatic progress with a reconceptualization of the etiology of their symptom complexes. With a new understanding of the physiological contributions in attachment-disordered children, psychopharmacological intervention has contributed substantial improvement in large numbers of maltreated (abused and/or neglected) children with associated attachment difficulties.

This chapter describes, from substantial clinical experience, a new conceptualization of the role that mood disorders, specifically Bipolar Disorder, play in the lives of maltreated infants, toddlers, and young children who subsequently develop associated emotional and behavioral problems related to bonding and attachment. This conceptualization takes both trauma and attachment theories into account and is not in opposition to them, but offers different perspectives in reaching certain conclusions.

Diagnosing mood disorders in children and adolescents with attachment difficulties is a very gratifying experience because it is so rewarding to effectively treat such a responsive disorder. Preschool children, school-age children, adolescents, and adults with mood disorders are substantially underdiagnosed and misdiagnosed in every community.

An underidentification and underdiagnosis of mood disorders leads to inadequate and ineffective treatment plans for attachment-disordered children. There is a strong tendency within mental health professionals, and the medical profession in general, to overidentify anxiety disorders, psychological factors, and particularly, Attention Deficit Disorder (ADD). In more than 25 years as a child psychiatrist, I have had hundreds of referrals of children with and without attachment problems with the specific working diagnosis of Attention Deficit Disorder on the part of the referring person or agency. Some of these children have indeed had ADD. Many other of these children have been misdiagnosed and in fact have mood disorders that require different treatment strategies and interventions.

Historically, mental health professionals have long associated Attention Deficit Disorder with Reactive Attachment Disorder. Children who have been maltreated do have attentional problems for many reasons. Common examples would include brain developmental and maturational factors, hyperarousal and hypervigilance secondary to trauma, and Attention Deficit Disorder. Some experts have put this correlation between ADD and Reactive Attachment Disorder at between 40 and 70% for maltreated, adopted children.

In my experience as psychiatric consultant to residential care centers, attachment centers, and adoption agencies as well as within my own private practice in which I supervise psychotherapists who work with children with attachment disorders, I have come to realize that Attention Deficit Disorder is overdiagnosed in this clinical population, leading to ineffective, even contraindicated, medical treatments. I have concluded that correlations between Bipolar Disorder and Reactive Attachment Disorder, Disinhibited Type, are much more common. This conclusion has led to more effective medical treatment plans for these children, resulting in greater emotional accessibility and receptivity, social reciprocity, behavioral self-control, and improved mood and self-esteem. An effective medical treatment plan helps these children act in more loving and lovable ways.

I have reached this conclusion about the comorbidity between Reactive Attachment Disorder and childhood Bipolar Disorder gradually over the past several years. In 1993 and parts of 1994, I was still diagnosing most attachment-disordered children with coexistent Attention Deficit Disorder. Since graduation from medical school, I am estimating that I have diagnosed and treated approximately 3000 cases of ADHD and 1500 cases of Bipolar Disorder. During this period, I have treated or consulted on an estimated 1000 cases of Reactive Attachment Disorder.

Three hundred children with Bipolar Disorder have been diagnosed and treated since 1994. Of this number, 100 children have had coexistent Reactive Attachment Disorder, Disinhibited Type. The 200 children with "only" a Bipolar diagnosis did not have histories of abuse or neglect.

This chapter is not research from a strictly scientific perspective. Clinical observation has commonly been the precursor to data-based medical research. History is full of clinical observation predating scientific documentation and validation. I have been fortunate to practice child psychiatry in a professional community whose focus has been on disruptive behavioral disorders associated with early life abuse and neglect. This community has attracted national attention and it has been my professional privilege and pleasure to assess and treat children from all over the United States and several foreign countries.

Psychiatric Diagnoses of Abusive Parents Are Indicators of Genetic Vulnerability for the Development of Mood Disorders in Maltreated Children

Diagnoses are formulated by evaluating three separate criteria. These are the child's personal and family history, emotional and behavioral symptoms, and mental status, the latter being an assessment of his or her current functional ability. An examination of these criteria is important to explain how childhood mood disorders correlate with Reactive Attachment Disorder.

In an attempt to understand physiological predispositions to various mental illnesses, a family history of mental illness is extremely important. Regrettably, mental health professionals, social service agencies, adoption agencies, and adoptive parents have had poor, even misleading, histories of birth parents of maltreated and subsequently adopted children.

Personality characteristics of abusive parents or other adult abusers have been studied. Not all adult abusers will demonstrate all of the following characteristics but many will exhibit several of these. Adult abusers commonly manifest both low self-esteem and poor impulse control with an external locus of control. In other words, they have relied on others, commonly authority figures, to limit their own excessive behavior. Paradoxically, most of them resent external control over their own behavior; they resist it and rebel against it. They have histories consistent with mood disorders, including rapidly changing and unpredictable mood swings. Inability to cope with stress, irritability, depression, and anxiety are common characteristics of abusive adults. They exhibit histories of antisocial behavior, including alcohol and substance abuse, as well as aggression toward others. Adult abusers lack social support and have histories of disrupted social relationships. Commonly, they have a history of abuse or neglect themselves as children. Abusive adults are extremely self-centered and seek power and control and take pleasure in being hurtful to others. Their sexual victimization of others can have a sadistic quality. Adult abusers display negative attitudes about children's behavior and lack empathy, knowledge, and understanding of normal childhood development and behavior. Demographic factors associated with adult abusers include poverty and poor education (not low intelligence). They live with or around high numbers of children.

I am unaware of any data extrapolating these personality characteristics of adult abusers into diagnostic entities of mental illness. Despite an inadequacy of scientific data regarding birth parents or other adult abusers, enough information has emerged so that it can be said with some degree of professional certainty that there are only a small group of birth parents who are capable of severely maltreating their children in an ongoing manner sufficient enough to create severe attachment problems. Among these diagnostic entities are disorders of psychological and biochemical cause as well as alcohol and substance abuse. Attention Deficit Disorder is not among these disorders.

Antisocial (Sociopathic) Personality Disorder

Many of the characteristics of children with Reactive Attachment Disorder also fit adult characteristics of Antisocial Personality Disorder. These include substantial conduct disorders, including cruelty to people or animals, lying, cheating, stealing, fire-setting, failure to conform to social norms with respect to lawful behavior, irritability, impulsivity, and aggressivity. These people have little regard for the truth and lack empathy and remorse. Many of these adults were themselves abused or neglected in early childhood.

Borderline Personality Disorder

The cause of Borderline Personality Disorder is not well understood, but there is evidence of both genetic and psychological influences, to some degree attributable to poor parenting (neglect or overprotection) between birth and 3 years of age. Borderline Personality Disorder manifests as long-term patterns of unstable mood, dysfunctional interpersonal relationships, and poor self-image. Their relationships are often characterized by alternating between extremes of idealization and devaluation of others. They exhibit identity disturbances, showing marked, persistent, and unstable self-image. They may be impulsive in ways that are potentially self-damaging;

for example, overspending, promiscuous sex, substance abuse, reckless driving, or binge-eating. They may exhibit recurrent suicidal gestures or threats or self-mutilating behavior. Their moods are notoriously unstable and reactive to marked degrees. They may experience chronic feelings of emptiness and display inappropriate intense anger or difficulty controlling anger. They are capable of transient stress-related paranoid ideation or severe dissociative symptoms. There is a growing trend among mental health diagnosticians to conceptualize Borderline Personality Disorder as a mood disorder, of genetic cause, treatable with mood-stabilizing medications.

Paranoid Schizophrenia

This complex disorder is present in under 1% of the general population. It is strongly genetically influenced, and is characterized by thought disturbances, such as delusions, with themes of distrust and hallucinations, usually auditory in nature. People with this disorder may be apathetic or have inappropriate affect or feeling tone. They tend to function at low levels of self-care, relate poorly to others, and have problems with emotional closeness or intimacy. Others have a difficult time getting close to them. As such, they do not frequently cohabit, form lasting intimate relationships, or have children. In a delusional or command hallucinatory state, they are capable of abuse or neglect, though this is uncommon.

Alcohol or Substance Abuse

In my experience working with abused children, this is the single most common characteristic of abusing parents or other abusive adults. However, it is also commonly a coexistent factor of abuse. While alcohol- and substance-abusing parents may severely abuse their children in an ongoing manner, purely alcohol- or substance-abusing parents with no other comorbid condition who overindulge or otherwise maltreat their children are ordinarily regretful and remorseful of their actions. They desire to stop but are unable to stop because of their addiction.

On the other hand, if alcohol- or substance-abusing parents also have coexistent Antisocial Personality Disorder, Borderline Personality Disorder, Paranoid Schizophrenia, or Bipolar Disorder, the intensity of the abuse is more severe and the extent of the abuse is far more lasting. Also, given the above comorbid factors, little remorse or neglect is felt, leading to a cycle of continuing abusive situations. Multiple studies demonstrate up to 60% correlations between alcohol and drug abuse with Bipolar Disorder. This is not to say that all drug- or alcohol-abusing people have Bipolar Disorder, only that these disorders commonly coexist within the same individuals.

Bipolar Disorder

Bipolar Disorder is a common mood disorder representing 1–3% of the general population. It is a genetic, inherited, familial disorder that ultimately results in biochemical imbalances within one's central nervous system. It manifests in manic (or hypomanic, a lesser form of manic) and depressive mood disturbances. In my professional experience, it is by far the genetic disorder that has the greatest coincidence with abuse or neglect of children and, as such, is the genetic disorder that children with coexistent Reactive Attachment Disorder of the Disinhibited Type also are predisposed to inherit. The degree of self-centeredness, irritability, and intensity of rage reactions while in a manic state is frequently sufficient to create severe maltreatment of children. Conversely, the degree of profound depression may also be likewise severe and prolonged enough to create longstanding maltreatment circumstances. In my experience, abusive parents with mild to moderate degrees of Bipolar Disorder would not tend to sustain abusive behavior to cause an attachment disorder. On the other hand, birth parents with a severe form of this disorder, particularly if this also includes psychotic features, contribute to enhanced potentials for severe and ongoing abusive circumstances, so that attachment difficulties are more likely.

There have been various epidemiological studies investigating the prevalance of Bipolar Disorder. These studies have been done before the inclusion of Bipolar II Disorder. Bipolar II Disorder manifests as recurrent major depressive episodes with occasional interspersed hypomanic episodes. There are also large numbers of alcohol- and substance-abusers whose underlying mood disorders go unrecognized, undiagnosed, or untreated. Drugs and alcohol are frequently the chosen "medication" of mood-disordered people. Many abusive parents, if prosecuted at all, are dealt with exclusively within the criminal justice system and are not evaluated for mental illness. It is my professional opinion that large numbers of abusive birth parents possess forms of mental illness, particularly Bipolar Disorder, that are never diagnosed or treated.

Statistics also point toward high incidences of Attention Deficit Disorder within prison populations. At its core, since Attention Deficit Disorder manifests only as inattention, impulsivity, and hyperactivity, it seems that attentional disorders are overrated as a factor leading to sustained, severe conduct disturbances and that prison populations are full of people with undiagnosed Bipolar Disorder. All of these factors lead to underrecognitions of the prevalance of this disorder in our communities.

Attention Deficit-Disordered birth parents uncommonly, even rarely, manifest sufficient self-centeredness, irritability, or intensity and frequency of rage reactions to create or sustain abusive circumstances. Attention Deficit-Disordered parents are reality based, generally have a high regard for their children, and even if they were to uncommonly abuse their children, have enough regret and remorse so as to learn from their own experiences and not repeat such actions. The abuse that they might ordinarily render is of a more mild or transient nature and not of the severe or prolonged degree that is associated with children with emotional attachment or bonding problems.

Remorse and regret are among the greatest motivators for change. Without remorse, positive change rarely occurs. Remorse is a growing distress within the individual arising from a sense of guilt over past wrongs. Remorse indicates the abuser has compassion for how victims feel. With remorse they are less prone to repeat their actions.

Schizoaffective Disorder

Schizoaffective Disorder is a genetically inherited disorder that affects biochemical integrity within the central nervous system. The essential features are an uninterrupted period of illness during which there are both mood (of a depressive, manic, or mixed nature) and cognitive symptoms, such as delusions or hallucinations. People with this disorder respond to similar medical treatment plans for the treatment of Bipolar Disorder. Within this chapter, further reference to Bipolar Disorder could be inferred as applicable as well to Schizoaffective Disorder.

Adaptation

Adaptation is the most common unifying characteristic of all living things. Adaptation is the characteristic through which all living things survive. The pivotal concept in the theory of evolution is adaptation. Evolution's emphasis on adaptation is the part played by motivation and control in an organism's interaction with its environment. It concerns pain and pleasure, emotion and thought, and desire and action, utilizing them in an adaptational framework. In biology and psychology, adaptation means change within the organism, calculated to equip it better for survival in various environmental conditions.

Given the environmental conditions of abuse and neglect, it is important to seek information from both trauma and attachment theories. Reactive Attachment Disorder is, at its essence, a Posttraumatic Stress Disorder of infancy and toddlerhood. In trauma theory, a persistent atmosphere of fear leads to vigilance, compliance, withdrawal, avoidance of stimuli associated with a known trauma, or a numbing of general responsiveness. A persistent atmosphere of fear also leads to ongoing symptoms of increased arousal physiologically, such as increased heart rate, startle responses, and sleep disturbances. Only when in actual terror, more than generalized arousal, does there tend to be defiance, resistance, or opposition. In my experience, passive resistance and opposition are far more adaptive to the survival of abused children than overt resistance or opposition. Fear, hypervigilance, emotional withdrawal, and behavioral compliance are more common adaptations to severe abusive circumstances in infants, toddlers, and young children.

In attachment theory, the inability to securely bond leads to anxious, ambivalent, avoidant, or confused attachments, with emotional and behavioral consequences being anxiety, and distrust of oneself and others. These children display depressed moods with helpless and hopeless aspects. Attachment-disordered children have internalized beliefs that they are unlovable, rejectable, and abandonable, and they have little trust that their world is safe. Anger would commonly be self-directed or displaced and projected onto innocent other people.

Psychologically, overt opposition and defiance are not positive adaptations to abuse or neglect. It is my experience that "loyalty conflicts" do exist within some children between abusive adults and their foster or adoptive parents. "Identification with the aggressor" can occur. These children do have a negative self-image, and are regressed and fixated with motives of self protection. As such, they distort reality and perceive danger where it does not exist. They can become aggressive as a result of these perceptual distortions. It is also documented that traumatized children can exhibit brain maturational delays. Subsequently, Disinhibition Syndromes associated with impulsivity and aggressivity can occur within these same children. Despite these psychological and trauma-based influences, it remains my clinical experience that the inherited, genetic predispositions to mental illness commonly outweigh these other factors. In the treatment of these children, if resistance and opposition are present and do not improve with appropriate treatment based on psychological or trauma theories, a consideration of another disorder, namely a mood disorder, should be given.

With persistent or sustained trauma and its consequent attachment-disordered components, what would be most expectable is to observe children who are passively controlling. Common personality characteristics would include sneakiness, underhandedness, deviousness, obstinacy, procrastination, emotional distance or emptiness, passive-aggressivity, or other covert actions. Conning and cunning would be adaptive mechanisms to deal with abuse or neglect. Indeed, such children are well known and, in essence, fit under the category of Reactive Attachment Disorder, Inhibited Type. Their emotional and behavioral inhibition demonstrates appropriate and understandable adaptation to their maltreatment.

Within clinical practice, a majority of the children identified with Reactive Attachment Disorder are disinhibited and exhibit overt opposition and defiance. Some of these children even display rage reactions, including verbal and physical aggressivity. It is these cluster of symptoms, among others, that are not positively adaptive for survival under circumstances of abuse, neglect, or other maltreatment. It is these maladaptive cluster of symptoms in which something else is occurring—Bipolar Disorder. A prominent subgroup of children with Reactive Attachment Disorder have inherited a genetic predisposition to Bipolar Disorder and the trauma they have experienced acts as a stress trigger to unleash overt symptoms of opposition, aggression, and sustained rage. I would estimate the coexistance of Bipolar Disorder with Reactive Attachment Disorder as occurring in 75–80% of all cases of the disinhibited type. It is this same cluster of symptoms that mental health professionals have included under the concept of Attention Deficit Disorder.

Misunderstandings about ADD and Bipolar Disorder

There is an overwhelming misunderstanding among most social service, educational, psychotherapeutic, and medical professionals concerning child and adolescent Bipolar Disorder. This disorder is commonly misdiagnosed as Attention Deficit Disorder. This chapter hopes to stimulate an increased awareness in understanding the similarities and differences between Attention Deficit Disorder, child Bipolar Disorder, and Reactive Attachment Disorder, Disinhibited Type. Differentiating these disorders leads to vastly refined and more effective medical treatment plans for these children. Specifically, children with Bipolar Disorder will be treated with more effective medications that treat severe and different biochemical imbalances within the Central Nervous System. Diagnosing comorbid Reactive Attachment Disorder and Bipolar Disorder in severely emotionally and behaviorally disturbed children and adolescents is important. When these comorbid conditions are recognized together, medications are able to improve emotional accessibility, receptiveness and reciprocity, self-control, mood, and self-esteem. These children are capable of making dramatic positive gains. These gains also make it far easier for other caregivers, be they parental, social service, educational, psychotherapeutic, or medical, to develop treatment plans that contribute toward more optimal effects.

"I can relate to this child so much easier now!" is a statement I have been fortunate to hear often. Each of the members of the treatment team feels more potent themselves at being able to make positive contact and communication with these children who are otherwise very difficult to communicate with. I have heard from speech and language therapists that their attachment-disordered children, once they have been additionally diagnosed and treated for a mood disorder, are so much more connected and motivated to improve. Massage therapists working with these children have noted their clients' improvements in body tension, defensiveness, and armour. Teachers and educational psychologists note progress in their functional ability and cooperativeness, and improved mood and motivation of their attachment-disordered students. Psychotherapists comment that these children are emotionally more engagable and capable of forming improved social relationships. An effective medical treatment for childhood Bipolar Disorder, along with treating the attachment component, enhances the quality of psychotherapy and other interventions to dramatic degrees.

A recognition of any mood disorder, but most notably Bipolar Disorder, brings about a recognition of a true medical disability. Information given to the various members of the child's treatment team regarding the genetic and biochemical aspects of this disorder enable them to be more empathic, understanding, encouraging, and supportive of the child. Commonly, Reactive Attachment-Disordered children are on the unlikable side. Having been maltreated, their confidence in themselves and their trust in other people results in insincere, phony, controlling relationships. An effective medical treatment has a stabilizing effect on their mood. These children are "in a better mood" with resultant improvements in many aspects of their lives.

Other than the child, the people who otherwise are most positively affected are the adoptive child's parents and siblings. Attachment-disordered children avoid and disrupt intimacy. Resolutions are made difficult because the child consistently stirs up unresolved conflicts. The inclusion of an attachment-disordered child to an otherwise functional family can influence disruptions in all family members. The family dynamics that had become altered in a negative way with the addition of an attachment-disordered child become more flexible with an effective medical treatment plan. With the recognition of the comorbidity between attachment disorders and childhood Bipolar Disorder, there is always improved family functioning with the inclusion of an effective medical treatment plan for these comorbid children.

"Severe ADHD May Precede Teenage Bipolar Disorder." This is one of the headlines on page 1 of the *Clinical Psychiatry News* in its April, 1998 edition. This article states that in many patients diagnosed with mixed Bipolar illness during adolescence, a characteristic history can be traced back to Attentional Deficit Hyperactivity Disorder beginning in early childhood, commonly as early as 3–4 years of age. The article goes on to mention research by Dr. Joseph Biederman, suggesting that between 1/5th and 1/4th of children with Attention Deficit Hyperactivity Disorder have Bipolar Disorder or go on to develop it. Because of their complexity, they are commonly additionally diagnosed as Oppositional Defiant Disorder, Generalized Anxiety Disorder, Separation Anxiety Disorder, and Obsessive Compulsive Disorder. It also quotes Dr. Joshua Feder, a psychiatrist, as noting that up to 60% of adolescents with mixed Bipolar Disorder develop substance abuse problems. This article suggests, and I strongly disagree, that the true precursor of adolescent Bipolar Disorder is "severe ADHD." The article concludes that it is when these children reach puberty that the more bipolar nature of their illness becomes more apparent. Extended information is provided within this chapter, outlining similar and different symptom complexes to facilitate a differentiation between these two apparently confusing disorders.

There is also a trend toward a recognition and treatment of comorbid

Attention Deficit Disorder and Bipolar Disorder of Childhood. There are recognized children with principally attentional problems who also have mood components to their symptomatology. Similarly, there are mood-disordered children who display prominent attentional difficulties as a part of their symptom complex. There are some concerns regarding this comorbidity. The general awareness of ADHD so far surpasses awareness of childhood Bipolar Disorder that mental health professionals are still prone to think of this primarily along attentional lines. I have also witnessed how stimulant medications, the most common medications for ADHD, can worsen already arousable and excitable aspects in these children.

It is helpful to conceive of this comorbidity as a ratio of mood-disordered versus attentional symptoms and to recognize that, at times, attentional symptoms are improved with stabilization and improvement in mood functioning. It is also common for mood-disordered children to have motivational problems that are misunderstood as attentional problems. As such, the occasional addition of an antidepressant medication (such as bupropion) instead of a stimulant medication enhances overall mood and improves attentional symptoms.

There are a number of understandable reasons for the overidentification and overemphasis of Attention Deficit Disorder and an underrecognition and underemphasis of Bipolar Disorder. Within the past few to several years, a massive amount of media attention has focused on ADHD. There are many conferences, articles, books, television, and movie portrayals that emphasize ADHD. There is also a stigma associated with the diagnosis of Bipolar Disorder. It is understandable that some professionals would prefer to initially emphasize a more innocuous disorder. Many professionals either are unaware of Bipolar Disorder as affecting children or are unschooled in their recognition of it. Childhood Bipolar Disorder is not emphasized in professional schools of education, psychology, medicine, or even in postgraduate child psychology or child psychiatry fellowship programs. In its natural state separate from attachment disorders, Bipolar Disorder in children may not be recognized because the parents, themselves, may have this same disorder and do not recognize symptoms as pathologic. These parents, commonly underdiagnosed themselves, fail to recognize the implications of their children's behaviors. Another factor is that prominent national authors in well-regarded texts include common symptoms of ADHD to include sleep disturbances, irritability, oppositional and defiant tendencies, rage reactions, and mood lability. While there is no dispute that these symptoms can and do occur in ADHD children, it is my experience that they have been overemphasized and contribute to confusion among professionals. This would be particularly true when relating severe symptoms associated with Reactive Attachment Disorder. All of the above-noted symptoms are common to Bipolar Disorder, with or without concomitant attachment difficulties.

A further complicating factor is that some bipolar children initially appear to favorably respond to stimulant medications. Some bipolar children, because they are inherently distractible, do demonstrate improvements in their concentration and attention with stimulant medications. However, stimulant medications are usually a mixed blessing. While they commonly can help concentration and attention, over the long run, stimulants can exacerbate arousal symptoms in an already arousable child. It is like saying to those of us with anxiety but without either Attention Deficit Disorder or a mood disorder, "You're anxious, go drink a pot of coffee!" If professionals are aware of seeing children or adolescents with a diagnosis of ADHD who are on stimulant medications, who have not already made sustained and substantial progress with their stimulant medications, and who have the above symptom complex (mood lability, unsatisfiability, oversensitivity, overreactivity, irritability, oppositional and defiant tendencies, temper outbursts and rage reactions), what is being observed is probably not ADHD. A serious reconsideration of a different diagnosis should be undertaken.

There are some encouraging signs that some mental health professionals are increasingly recognizing and treating childhood Bipolar Disorder effectively and appropriately. An example is a paper presented at the American Psychiatric Association Annual Convention in May of 1998 entitled, "Treatment of Bipolar Spectrum Symptoms in Children and Adolescents: A Retrospective Multi-Study." This paper was authored by Ronald Rabin, M.D., and Clifford Siegel, M.D., and reported on retrospective studies of 194 children, ages 3–17. Patient's charts were reviewed on all children treated for bipolar spectrum symptoms whose current medication regime was of at least 6 months' duration. On initial assessment, 88% of patients demonstrated at least three of six presenting mood symptoms. These included mood lability, irritability, rage reactions, hyperexcitement, and depressed or elated moods. The most prominent associated symptoms included sleep disturbances (80%), concentration and attention problems (69%), suicidal ideation (56%), defiance and conduct problems (52%), anxiety states (49%), psychotic symptoms including hallucinations and/or delusions (46%), severe risk-taking (23%), suicide attempts (22%), substance abuse (20%), hypersexual behaviors (20%), and obsessive compulsive symptoms (10%).

The duration of the symptoms in these patients was quite long, commonly greater than 5 years in most patients. Fifty percent of the patients had demonstrated significant symptoms prior to age 6. Family history was positive for Bipolar Disorder in 47% of the cases, positive for depression in 87% of cases, and positive for substance abuse in 43% of cases. There were 22 completed suicides among extended family members. Fifty-eight percent of the group had prior medication treatment, primarily involving stimulants or antidepressants. Eight-one cases had prior stimulant treatment. Among those, 26% reported that these stimulants had been ineffective and 44% were reportedly made worse with destabilization of mood and the appearance of increased aggressivity.

The experience of the group with antidepressants was also significant. Seventy-two cases had prior trials of antidepressants. Of these, 22% reported that the antidepressant medications were ineffective, while 62% reported that symptoms were made worse with increased agitation and aggression. Treatment with mood-stabilizing medications was found to have been quite effective. The vast majority of the patients were treated with either valproate or lithium. Several patients had remission of mood and associated symptoms with mood stabilizers alone, while some patients required additions of other medications. Ten percent of the population demonstrated only limited response and remained significantly unstable.

In my own experience since 1994, in the 300 children and adolescents treated for Bipolar Disorder, the concept of "mixed type" of Bipolar Disorder has been the most prominent presentation that these children have. A mixed type of Bipolar Disorder is also commonly referred to as "rapid cycling." In this type of illness, there are features of dysphoria manifested by symptoms of unsatisfiability and unpleasability. Nothing is good enough for these children. The remainder of this mixed type of presentation represents hypomanic features consisting of grandiose mood, irritability, opposition, temper outbursts, and occasional rage reactions and aggressivity. There is some "bell shaped curve" to the presentation of these children. As such, there are some children who present with almost no depressive symptoms ever, presenting with almost exclusive hypomanic components. Conversely, there are other children whose unsatisfiability and unpleasability extend into periods of persistent low mood which, upon occasion, can even express itself in suicidal ideation, with or without intention. This dysphoric, depressive side to these children is often mistakenly diagnosed as Dysthymic Disorder or Major Depressive Disorder. Subsequently, these children are commonly placed on antidepressant medications, which have poor effects.

Could the Correlations with Reactive Attachment Disorder, Disinhibited Type, Be with Conduct Disorder, Oppositional Defiant Disorder, or Posttraumatic Stress Disorder, Not with Bipolar Disorder?

Ten to 20% of children with disruptive behavioral disorders associated with early life abuse or neglect do not respond well to any psychoactive medications. These are the percentage of children with the worse prognosis. Not only does medication not work but neither do other treatment modalities. These are the children prone to chronic conflicts with authority. They grow up to have continued antisocial traits and do poorly, often winding up in penal institutions. In my experience, these children demonstrate symptoms consistent with Conduct Disorder, Oppositional Defiant Disorder, maturational brain abnormalities associated with abuse, or the inhibited form of Reactive Attachment Disorder in which they adapt by conning and cunning exclusively.

Conduct Disorder is a repetitive and persistent pattern of behavior in which the basic rights of others or major age-appropriate societal norms are violated. These behaviors fall into four main groupings: aggressive conduct that causes or threatens physical harm to other people or animals; nonaggressive conduct that causes property loss or damage; deceitfulness or theft; and serious violation of rules. The behavior pattern is usually present in a variety of settings, such as home, school, or the community.

Oppositional Defiant Disorder is negativistic, defiant, disobedient, and hostile behavior toward authority figures that is characterized by frequent occurrence of at least four of the following behaviors: losing temper; arguing with adults; actively defying, refusing to comply with the requests or rules of adults; deliberately doing things that will annoy other people; blaming others for their own mistakes or behavior; being touchy or easily annoyed by others; being angry, resentful, spiteful, or vindictive. As is also true in the previous discussion regarding Borderline Personality Disorder, there are studies demonstrating the effectiveness of mood stabilizers in the medical treatment of Oppositional Defiant Disorder.

There is substantial overlap between the symptoms within either Conduct Disorder, Oppositional Defiant Disorder, or Bipolar Disorder. It is important to look at subtle differences between these three disorders. Despite Jim's case history, it is a mistake to believe that bipolar children's moods fluctuate in cycles. It is my clinical experience that the vast overwhelming majority of children with Bipolar Disorder present in a "mixed fashion," that they have both hypomanic and dysphoric components commonly at the same time. Children with strictly Conduct Disorder or Oppositional Defiant Disorder do not usually display dysphoria as a prominent symptom. The mood within bipolar children is one of common grandiosity and subsequent entitlement, believing that they are deserving of special favors without needing to work for them. Children with Conduct Disorder or Oppositional Defiant Disorder by themselves do not tend to display grandiosity or entitlement. Mood-disordered children are often (not always) verbally articulate and display unique creative abilities. Children with Conduct Disorder and oppositional tendencies commonly do not. Children with Bipolar Disorder are commonly emotionally wired, tense, anxious, and agitated. Sleep disturbances are far more common in mood-disordered children than in children with Conduct Disorder or Oppositional Defiant Disorder.

Most importantly, the reasons why either Conduct Disorder or Oppositional Defiant Disorder exclusively do not correlate with Reactive Attachment Disorder are found in discussion of adaptation. In consulting about these children since 1977, it is my clear experience that the overwhelming emotion affecting abused, neglected, and severely maltreated infants, toddlers, and young children is fear. Fear begets physiologic hyperarousal, hypervigilance, and also psychological withdrawal, avoidance, low self-esteem, and behavioral overcompliance, not overt opposition and rage. Overt opposition and rage are not positively adaptive under abusive circumstances. Therefore, a return to the concept of mood disorder, especially Bipolar Disorder, is warranted.

These disorders not only overlap in symptoms, but they also are not mutually exclusive diagnostically. It is very common for me, in evaluating these children, with disruptive behavioral disorders secondary to early life abuse, neglect, or substantial mistreatment, to diagnose all of the following comorbid conditions within the same individual. These include: Oppositional Defiant Disorder (or Conduct Disorder); Posttraumatic Stress Disorder; Reactive Attachment Disorder, Disinhibited Type; and Bipolar Disorder, Mixed Type, Severe (with or without transient psychotic features).

It is my professional opinion that all of these children should be considered first to have some form of mood disorder, most specifically Bipolar Disorder. Being symptom based, the implications for causes of both Conduct Disorder and Oppositional Defiant Disorder are not biochemical, at least not clearly defined biochemical causes. The standard treatment for Posttraumatic Stress Disorder commonly includes antidepressants, which can exacerbate arousal tendencies. With the assumption of a mood disorder, the prognosis regarding treatment of these children actually dramatically improves. The diagnosis of a mood disorder in these children improves their overall prognosis for treatment because medications for Bipolar Disorder are capable of producing moderate to substantial improvement in the overall symptom complex. Assuming only Oppositional Defiant Disorder or Conduct Disorder without any underlying mood component assumes either psychotherapy and/or behavioral treatments primarily or the use of medications that may not cover the spectrum consistent with mood disorders.

Correlations between Dysthymia or Major Depression and Reactive Attachment Disorder, Inhibited Type

Case History

Ekatrina was an 11-year-old adopted child from an Eastern European country. She had been adopted, along with her birth sibling, for 5 years. She was referred for a psychiatric evaluation by her adoptive parents, attachment therapist, and therapeutic foster parents. Ekatrina had been referred from her home state following intense treatment by a well-regarded attachment therapist. This intense treatment was over a 4-year period with minimal to no gains made. She displayed the classic attachment disorder symptoms of lack of eye contact and phoniness, with little conscience or remorse over her activities. She was described as exceedingly passively controlling with literally everyone in her environment, emotionally empty, and having limited access to her own feelings. She had an extremely negative self-image and substantial periods of "emotional shutdown." Periods of "voluntary mutism" were described in which Ekatrina would not relate to others for hours and sometimes for days. Her passive-aggressive control was seen both as a manifestation of her low self-esteem and her attempt to control and manipulate others. She was described as a "black hole sucking everything around into it." In contrast to her shutdown, periods of "agitation" were also described. During those times, she was described as fidgeting, pulling her fingers, rubbing body parts, putting her hands in her mouth, eating her hair, and continuing to adapt poorly during the 5 years she had been with her adoptive parents. There were also documented sensory processing disorders and fine motor and auditory processing deficits. Trials of methylphenidate (Ritalin) and guanfacine (Tenex) had been tried with "neutral" effects.

Historically, Ekatrina's birth mother was described as "seriously mentally ill with strong tendencies toward dissociation." Her birth father was described as "violent and alcoholic" and was imprisoned for the attempted murder of Ekatrina's mother and grandmother. There was some belief that she may have been sexually abused by her birth father. Following removal from her birth family and placement in an orphanage, there was evidence of both physical and sexual abuse prior to her adoption.

Following consultation and evaluation, the following psychiatric diagnoses were made. In addition to her multiple learning disabilities and processing disorders, a diagnosis of Reactive Attachment Disorder, Inhibited Type (American Psychiatric Association, 1994) was provided. Almost all of Ekatrina's symptoms related to passive control of her environment. Her low trust levels, sneakiness, underhandedness, deviousness, conning, and cunning could be seen as understandable adaptive mechanisms to her early trauma, abuse, and neglect. Her chronic history of "feeling miserable and pitiful" with intermittent "poor appetite manifesting as hunger strikes" were a part of Ekatrina's symptom complex. Her long-standing symptoms of persistent low mood and low self-esteem, coupled with low energy and poor concentration indicated a diagnoses of Dysthymic Disorder, Early Onset (American Psychiatric Association; 1994).

Given the above symptom complexes and diagnoses, a clinical trial was recommended of Citalopram (Celexa). Ekatrina was begun on 20-mg tablets, a half-tablet each morning for 10 days, then one tablet each morning for 3 weeks. On follow-up 1 month later, she was described as being in a more consistent, positive mood, and exhibiting a "real voice and smile." She was described as exhibiting more genuine emotions and being more "workable" by her psychotherapist. Within a few weeks time, there was a deterioration of Ekatrina's mood and a return to her passive, resistant, controlling self. Recommendations at that time were made to increase the Citalopram from 20 to 30 mg for a few weeks and, if necessary, up to 40 mg daily.

Upon follow-up, this medication at this dose has contributed to a more sustained improvement in mood and sense of well-being for Ekatrina. While she remained seriously disturbed, she was more receptive and responsive to psychotherapeutic techniques than she had been previously. It appeared, though it was still early in her treatment, that her longer-term potential to be helped was brighter. A greater degree of hope and optimism was expressed by her psychotherapist, therapeutic foster mother, and adoptive parents.

As previously stated, it is my experience that the disinhibited form of Reactive Attachment Disorder commonly coexists with the comorbid state of Bipolar Disorder. The inhibited form of this disorder is a more appropriate psychophysiological adaptation to early life abuse and neglect. Given the severe psychopathology of birth parents' abuse or neglect of these children, there is a correlation of the inhibited form of this attachment disorder with either Dysthymic Disorder, which manifests as persistent low mood and motivation, or Major Depressive Disorder. These children would experience their trauma by psychologically and physiologically adapting in withdrawn, inhibited, insecure, anxious, and vigilant manners. It is my experience that this subgroup of Reactive Attachment Disordered children, who do not generally display disinhibition of a severe degree, do benefit from considerations of antidepressant medications. Antidepressant medications have long been useful for treating both Dysthymic Disorder and Major Depressive Disorder. There is also ample documentation that antidepressant medication may be useful in treating mood and hyperarousal symptoms associated with Posttraumatic Stress Disorder.

Implications for Adult Survivors of Childhood Physical and Sexual Abuse

I have accumulated 12 case histories since 1994 of adult patients, all women, who were severely physically, sexually, and emotionally abused as children by birth parents, siblings, and other birth ancestors. All of these women's abusive circumstances began in childhood after the age of 5. Several were not prominently abused until their early adolescence. In other words, the attachment component of their development was not a significant factor in these girls' lives.

All of these adult women had sought and obtained various forms of psychological and psychiatric treatment, most for several-year periods. All had diagnoses of chronic Posttraumatic Stress Disorder and Major Depression. All were treated with psychotherapy, antidepressants, and occasional antianxiety medications. None of these women were doing well emotionally or behaviorally. They were severely restricted in their activities. A few were completely disabled emotionally as a result of their experiences and were on Social Security Disability. Only a few of these women were working productively.

As explained earlier in this chapter, diagnoses result from personal and family histories, symptoms, and mental status. By evaluating the personality characteristics of their abusive birth parents, siblings, and other ancestors, it was postulated that these abusers suffered from severe mental illness, most likely Bipolar Disorder.

These women did display substantial symptoms of traumatization and depression. They were avoidant, withdrawn, isolative, and behaviorally overcompliant. They had low self-esteems and displayed poor trust and hypervigilance. The obvious assumption of their treatment providers was exclusively Posttraumatic Stress Disorder and Major Depressive Disorder, Recurrent Type.

None of the treatment providers seem to have taken genetic vulnerability and predisposition to Bipolar Disorder into consideration. No one had ever asked these women about common symptoms of Bipolar Disorder. Because of their shame and humiliation, these are not the type of patients who will volunteer the following symptoms. These symptoms included mood swings, periods of relative well-being, racing thoughts contributing to difficulty getting to sleep, alcohol or substance usage, sexual overactivity with poor judgment, oppositional tendencies, and even temper outbursts and rage reactions. The belief that this type of patient is consistently depressed is misleading. While they would complain of a depressed mood, it becomes important to ask if that depressed mood ever lifts; commonly it does for periods of time. These are subtle, yet important, differences between Major Depressive Disorder that tend to be consistent with its depressed mood and Bipolar Disorder that tend to be inconsistent with its depressed mood.

A reconceptualization of causal factors based on personal history, family history, and symptom complexes suggested the potential for genetic vulnerability and predisposition to Bipolar Disorder. Understanding each patient's causes of her problems as both mood related and trauma related became the key to solving her clinical puzzle. The trauma each of these women experienced added an enormous overlay of anxiety, agitation, and self-doubt. Their bipolar symptoms were masked and symptomatically overshadowed by their traumatic experiences. Like the children in this chapter described with coexistent Reactive Attachment Disorder and Bipolar Disorder, the understanding and effective medical treatment of a comorbid Posttraumatic Stress Disorder and Bipolar Disorder contributed greatly to these women's progress.

With this reconceptualization of a dual diagnosis of Posttraumatic Stress Disorder and Bipolar Disorder, multiple changes were able to be made in their medication regimes. All 12 women were given trials of mood-stabilizing medication with resultant improvement in all 12. Some of these women ended up on one mood-stabilizer alone, others on two complimentary and compatible mood-stabilizers, and others on mood-stabilizers and antidepressants. There were still others, because of the severity of their symptoms, who had an optimal response to mood-stabilizers and an atypical antipsychotic medication.

All 12 of these women had moderate to significant improvement in their emotional well-being, behavior, and functional ability. All had improved self-concept, intimacy, and social relationships. All of them were more substantially productive with their lives.

These retrospective case histories offer further corroboration that the theory expanded on for the correlation between Bipolar Disorder and Reactive Attachment Disorder in children can have extensions into adult life for those people severely abused after the developmental challenges of attachment and bonding have ended.

Trauma Theorists and Therapists Disregard Genetic Vulnerability in Maltreated Children with Attachment Disorders

Trauma experts emphasize exclusively the posttraumatic stress components of trauma and their impact on brain development, maturation, and function. They explain the diversity of symptoms in maltreated children as related to changes in immature central nervous systems. There is substantial scientific data to validate their concepts. There are some maltreated children who will most closely match these concepts. The effects of trauma on an immature brain can be enormous.

My concern is these theorists' minimization, even emphatic denial, that these children inherit predispositions of genetic vulnerabilities to mental illness. At an annual conference related to children's issues, held in Denver, Colorado, an international expert on the effects of trauma in children emphatically denied any genetic predisposition or vulnerability in abused, neglected, abandoned, or otherwise maltreated children, even if such abuse was perpetrated by birth parents or their direct ancestors.

Within this same 2-day conference, this international expert was asked about the psychopharmacological treatment of these children. In this expert's brief but clear reply, it was this expert's experience that both stimulant medications (for ADHD) and antidepressant medications had almost universally poor results. Lithium and valproate, both mood stabilizers, had by far produced the greatest benefits to these children.

I am in complete agreement with this expert's experience that the mood-stabilizing medications offer the best hope (along with the new atypical antipsychotic agents, which are expanded upon later in this chapter). I am in disagreement as to the reason why these same medications have the greatest effects.

It is my experience in some children with this comorbid condition of Bipolar Disorder and Reactive Attachment Disorder that the actual abuse or neglect they experienced was relatively mild—certainly not enough to promote severe brain maturational problems or explain the severity of their symptoms. Their birth parents' severe alcohol or substance abuse, and/or bizarre or antisocial behavior was enough to warrant removal from the home and subsequent relinquishment. Some were minor witnesses to violence rather than being perpetrated on themselves.

It is confusing why trauma theorists have to adhere emphatically to one exclusive explanation for the severity of these children's symptoms. We are in a toddlerhood ourselves at understanding the enormous complexities inherent within these children. My experience suggests multifactorial causes. A theory of genetic vulnerability to mental illness is not in opposition to trauma theory but expands upon it and offers different and, at times, more meaningful and valid explanations for these children's problems. Hopefully future study will elucidate the multiple causes, including both the effects of trauma on immature central nervous systems and genetic vulnerability and predisposition to severe mental illness.

CHARACTERISTICS OF ATTENTION DEFICIT DISORDER, BIPOLAR DISORDER, AND REACTIVE ATTACHMENT DISORDER, DISINHIBITED TYPE

In presenting these differential features, it is hoped the reader will gain sufficient information so as to more accurately differentiate these disorders. Please refer also to the outline at the end of this chapter.

Each of these disorders occurs in some spectrum. As such, there are interesting similarities, differences, and overlaps in the descriptions of these children. As such, one will see ADHD children who demonstrate mood component features and mood-disordered children who present with attentional or distractible features. Likewise, one will see children with attachment difficulties manifesting in mild, moderate, or severe forms. These disorders are not mutually exclusive.

Age of Onset

Retrospective histories with both ADHD and bipolar children will reveal that hyperactive and aggressive toddlers are very common. These children also are identified as special-needs children as early as preschool. By kindergarten and first grade, many of the ADHD children and bipolar children have been recognized for the symptoms they manifest. The onset of a more structured school environment accentuates each of these children's difficulties. Early adolescence, with its developmental and hormonal influences, also brings about a surge of recognition due to symptom initiation or exacerbation.

The age of onset for children with attachment difficulties is that early period in their lives associated with maltreatment.

Family History

Birth parents and other ancestors of children or adolescents with ADHD will show histories of attentional problems. Family history would show ancestors with academic difficulties, mostly based on difficulties with performance and task completion. A common ancestral history is one of alcohol or substance abuse.

The ancestors of bipolar children would have histories of mood disorders, most commonly depression, but there is an increased recognition of Bipolar Disorder. Academic or other performance difficulties would commonly focus more on distractibility, lack of motivation, or opposition and defiance with authority figures, including teachers. Histories of alcohol or substance abuse, often severe, would be included in the ancestral histories of bipolar children. It is also common in the family histories of bipolar children to have ancestors who have principally been hypomanic through much of their lives. These manifestations could result in ancestors who are extraordinary in their charisma, creativity, leadership, or accomplishments. Because of higher rates of adolescent sexuality among bipolar teenagers, there will be higher rates of unwanted pregnancies, relinquishments, and therefore histories of adoption within bipolar ancestors' histories. Also, given all of the previously referenced confusion that currently exists as well as the overlapping comorbidity between Attention Deficit Disorder and Bipolar Disorder, there will commonly be ancestors who will be identified as having ADHD.

The histories of birth parents and other ancestors of children with attachment problems will be replete with abuse, neglect, and other maltreatment. There will be very common histories of severe emotional and behavioral disorders among ancestors.

Prevalence

Several studies suggest that ADHD is represented in 3–6% of the child population. Different studies suggest that Bipolar Disorder exists in 1–3% of the general population, and there is evidence to suggest that this prevalence rate is similar in childhood populations. No clear statistics exist regarding the prevalence rate of Reactive Attachment Disorder. Its prevalence ranges from common to uncommon.

Etiology

The principle cause of Attention Deficit Disorder is genetic, although neurochemical, fetal developmental, brain traumas, and nutritional deficiencies would be additional factors in selected numbers of children and adolescents.

The etiology of Bipolar Disorder is genetic. This is an inherited, familial disorder that results in biochemical imbalances within selected parts of the Central Nervous System. While hormonal factors during puberty do not cause this disorder, their occurrence during the onset of puberty commonly brings on mood factors that had lain quiet during childhood.

Reactive Attachment Disorder is a complex psychophysiologic response to early life abuse, neglect, and abandonment.

Duration

Attention Deficit Disorder is a chronic and usually unremittingly continuous disorder that tends to improve with age. Individuals with this disorder commonly find positive ways to adapt and compensate. It can appear to worsen in early adolescence because of societal expectations regarding task completion and performance.

Bipolar Disorder may or may not show clear emotional and behavioral cycles. In children, it most commonly presents as a mixed mood disturbance with unsatisfiability and oppositional tendencies common. Bipolar Disorder tends to worsen over years with increased severity of symptoms. Hormonal factors worsen the severity of Bipolar Disorder. Hormonal factors will accentuate the inherent sexuality and aggressivity within the early teenage years.

Reactive Attachment Disorder's duration is dependent upon the intensity and extent of the abuse or neglect, the child's adaptation to that abuse, and the child's innate temperamental factors. Without adequate treatment, these children worsen over time and often develop antisocial character disorders.

Attention Span

It would not be called Attention Deficit Disorder for any more prominent symptom. Attention Deficit Hyperactivity-Disordered children's attention span leads them to think in a superficial manner, leading to lack of productivity and impulsivity. Several experts have put forth the concept of "hyperfocusing" in ADHD children. In my experience, this is an overrated phenomenon and is another factor that contributes to the misunderstanding and confusion between these disorders.

Bipolar children's attention span is entirely dependent upon their interest and motivation levels, which can fluctuate dramatically. Bipolar children experience boredom very easily, always seeking more pleasurable stimuli. This inherent distractibility in bipolar children is commonly mistaken for attentional problems. When motivated, bipolar children and adolescents are often capable of obsessively focusing for prolonged periods.

Children with Reactive Attachment Disorder are physiologically hyperaroused and hypervigilant, commonly scanning their environment in a self-protective intent to maintain their safety. This hypervigilance can mimic distractibility and does manifest as shortened periods of focus and attention.

Impulsivity

Impulsivity for most ADHD children is secondary to either inattention or obliviousness. Whatever dangerousness they may exhibit is ordinarily met with both regret and remorse.

Bipolar children's impulsivity is a manifestation of their mood-driven behavior at the moment. This drivenness is an almost "irresistible impulse" given the intensity of emotions that bipolar children experience. Given the intensity of mood, Bipolar children commonly become genuinely confused when their actions are judged or criticized. This confusion is the result of the mood-drivenness of their actions and a coexistent perception that "it must be the thing for me to have done." Their grandiosity and intensity of mood does not lead them toward self-doubt. All of these factors lead toward thrill-seeking or danger-seeking behaviors with little remorse or regret associated with their behavior. Bipolar children are also capable of counterphobic behavior, actually moving toward a certain danger that normative people would avoid.

Because children with Reactive Attachment Disorder range from highly overreactive to highly controlled, it would appear that the motive of self-protection and self-preservation would be the influential factor. Because of their controlling nature, their reactions often appear more deliberate and seemingly more purposeful. However, because of their reality distortions as a result of their trauma, they manifest poor cause-and-effect thinking and are therefore capable of acting in highly impulsive ways. They display minimal remorse or regret.

Hyperactivity

Fifty percent of children with ADHD display hyperactivity. They are commonly fidgety, jittery, and disorganized in their hyperactivity. Their hyperactivity levels moderate by mid-adolescence. The vast majority of children with childhood Bipolar Disorder display hyperactive symptoms. Hyperactivity is also common among children with attachment difficulties.

Self-Esteem

The self-esteem of many attention deficit disordered children is low, being rooted in ongoing performance difficulties. They distrust that their intentions and actions produce the same effect. This self-doubt is commonly reinforced with authority figures' need to structure their activities. High correlations with learning disabilities (up to 50%) and sensory integrative disorders accentuate and reinforce their low self-esteem.

Most bipolar children's self-esteem is low because of their inherent lability and unpredictability of mood. On the other hand, their grandiose or expansive mood can commonly mimic a positive self-esteem.

The self-esteem of children with Reactive Attachment Disorder is low. This self-esteem is rooted in real experiences of physical and emotional abandonment. They feel worthless and unlovable. This unlovability is commonly masked by anger, emotional distance, or seeming indifference.

Mood

The mood of most ADHD children most of the time is friendly in a genuine manner. They can be short-tempered and irritable as self-protective anticipations of criticism or adverse judgment.

Bipolar children's moods are highly unpredictable and can range from overexuberant to gloomy and unpleasant, commonly possessing an attitude of dissatisfiability and unpleasability. Their exuberance is a manifestation of their expansive or even grandiose moods. Conversely, there are some bipolar children in adolescence who have prominent clear depressive moods, even manifesting in suicidal ideation or intention.

Attachment-disordered children's moods are commonly hard to gauge because of their self-protective, superficially charming demeanor. Their general distrust of their environment creates symptoms of phoniness, insincerity, guardedness, and emotional distance. They are difficult children to get to know in a genuine fashion because of their need to remain nonintimate.

Control Issues

Attention Deficit Hyperactivity Disordered children and adolescents tend to desire to seek approval, getting in trouble mainly by their consistent inability to complete tasks. Their low self-esteem and desire for control commonly influence them to develop passive-aggressive personality styles in which they control others through dawdling, procrastination, and inadequacy. This psychological overlay can be difficult to differentiate from the attentional factors themselves. Caregivers are commonly put into a dilemma with passive control by ADHD children. If caregivers are too lax or unstructuring, they may experience their children as floundering or inadequate. On the other hand, if ADHD children perceive caregivers as overly judgmental or controlling, their passive-aggressive style would be a covert way of showing anger, hostility, passive control, and revenge.

Bipolar children may intermittently desire to please but they commonly push limits and relish power struggles. They commonly consider themselves "winning" if they are able to engage others in power struggles, not always if they win or lose. Bipolar children are commonly verbally articulate, and use their words in a persuasive and manipulative fashion. They are experts at engaging other people in power struggles.

Attachment-disordered children tend to be more behaviorally controlled but are commonly coercive as an adaption to a fear of losing control relative to their traumatic past. Their true controlling nature has a passive but exclusively self-serving purpose. Having low trust levels in others and not envisioning themselves as powerful, they gain their control through underhanded, covert, sneaky behavior.

Opposition and Defiance

While ADHD children demonstrate argumentativeness, there is nonetheless a tendency to relent with some show of authority. Having short attention spans, they do not dwell or focus on any one circumstance long enough to pursue in an argumentative or intense fashion. As such, this characteristic allows them to be far more redirectable by authority figures.

Bipolar children and adolescents are usually far more overtly and prominently defiant, often not relenting to authority. They are more relentless in their insistence on getting their own way. Some bipolar children will be exclusively passive-aggressive.

Children with attachment disturbances are conning and cunning. They are covertly defiant, showing their opposition and defiance in passive-aggressive manners.

Blaming

Children with Attention Deficit Disorder tend to blame others or external circumstances as a self-protective mechanism to avoid immediate adverse consequences. While they have difficulty sustaining responsibility, they are nonetheless usually accountable for their actions. They are able, however unwillingly, to make up for their misbehavior.

Influenced by some grandiosity of mood, children with Bipolar Disorder blame others or external circumstances out of a genuine disbelief or denial that they could have caused something to have gone wrong. This is because of their need to have cognitive congruence with their moods. It is as if they are saying to themselves, "Since what I feel is so intense, it must be the right thing for me to do." Bipolar children are difficult to hold responsible and accountable, unless it is under circumstances in which they are motivated or interested.

Children with attachment disorders perpetually perceive themselves as victims. Coming from a helpless position, they deny any power. They are rejecting of responsibility and accountability.

Lying

Children with Attention Deficit Disorder will lie in order to attempt to avoid immediate adverse consequences of their behavior.

Bipolar children will lie in a manipulative attempt to "get away with it." Their lies may be more elaborate and even believable. Lying is commonly a component of control issues: saying what one needs to say at the time to get what one wants.

Children with attachment difficulties exhibit "crazy lying," in other words, lying about the obvious. They are psychologically fixated in self-centered, self-protective, "primary process" primitive distortions. These perpetuate a "victim position" and passive control, of which lying is one common feature.

Fire-Setting

While children with Attention Deficit Disorder will commonly play with matches out of curiosity, their intentions are ordinarily not malicious and actual fire-setting is uncommon. Bipolar children have a much greater difficulty with limits and are intrigued with matches and fire as a form of dominance or overt control. As such, they are far more prone to not only curiosity but actual fire-setting. They are capable of malicious intention. If this is a prominent feature, it is indicative of either a severe form of Bipolar Disorder or other additional personality or character-disordered components.

Children with an attachment disorder will set fires out of a revenge motivation and can be malicious in their intention. Some of their danger-seeking behavior may be secondary to internal despair.

Irritability, Anger, Temper, and Rage

Of all the different characteristics outlined in this chapter, this particular characteristic commonly possesses the greatest differences in degree and intensity among the three different diagnoses. Most ADHD children's anger is situational and in response to overstimulation, poor frustration tolerance, and need for immediate gratification. Temper outbursts are usually short-lived, lasting several minutes and uncommonly up to an hour. There is often remorse and regret over excessive behavior. Destructiveness is secondary to overstimulation and therefore usually accidental.

The rage of children with Bipolar Disorder is commonly secondary to either limit-setting or attempts to control their excessive behavior. Their rages may be explosive and gone in a flash or can be of intense proportions and long-lasting. Children with Bipolar Disorder are capable of rages that can be sustained for up to a few to several hours duration. The amount of literal energy they are capable of expending in a rage reaction is extreme and difficult for anyone without Bipolar Disorder to simulate. An adult, on the other hand, could simulate the amount of anger in a child with Attention Deficit Disorder. Bipolar children, having more innate control issues, also demonstrate much more deliberate destructiveness.

Children with attachment difficulties tend toward chronic anger, or revenge as their motivation. Perceiving themselves as victims, they have retaliatory rationalizations for their destructiveness. They are hurtful to innocent and vulnerable younger children and pets.

Entitlement

Children with Attention Deficit Disorder may believe they are deserving of special benefits from time to time. Overgratification by their parents would accentuate this characteristic. Their need for immediate gratification is not necessarily the same as their believing that they deserve special treatment. As such, it is not a prominent symptom unless they are overindulged.

Children with Bipolar Disorder have both grandiose and expansive moods that contribute to the intensity with which they feel they deserve special treatment.

Attachment-disordered children may display some entitlement as a compensation for deprivation and abandonment. They are prone to hoard things from their experience of "not having enough." In my own experience, they act more undeserving than entitled.

Conscience Development

Children with Attention Deficit Disorder, raised in a reasonably normal environment, have good conscience development, showing remorse and regret over excessive impulsive behavior.

The conscience development of bipolar children is more limited and dependent upon their intensity of mood. As such, their actions are more mood driven than impulsive. Because of this mood intensity, there exists a cognitive congruency inherent with their mood. In other words, they commonly rationalize their behavior either internally or with others by saying something like, "If this felt so strong, it must have therefore been the right thing for me to have done." (See "Blaming.")

Children with attachment difficulties are "street smart" and tend to have good self-protective survival skills. They are devious con artists whose motive is self-protection. They can be calculating and deceptive.

Sensitivity

Since children with Attention Deficit Disorder have difficulty sustaining focus, they are commonly somewhat oblivious and unaware of the sets of circumstances they are in, and their immaturity and inappropriateness shows as a result. They get the "big picture" but miss the details. This can lead to a relatively "laid back attitude" in which they let go of conflicts relatively easily. They are capable of irritability and uncooperativeness, particularly in the early morning.

Bipolar children, in contrast to ADHD children, are acutely aware of circumstances, hypersensitive to how they are personally affected by those circumstances, and "hot reactors" to these circumstances. Children with Bipolar Disorder tend toward extremes of upset, even to obsessive degrees. They can compulsively and relentlessly "grind" on issues. They have high components of irritability and disagreeability, especially in the early morning.

To use a metaphor, the ADHD child will see the forest and miss the trees, whereas the bipolar child will be acutely aware of individual details, seeing the trees but missing the forest.

Compensating for past helplessness, attachment-disordered children are hypervigilant, emotionally and behaviorally resistant, and insensitive to the feelings or actions of others. They are rarely physically ill and have limited emotional repertoires.

Perception

Children with Attention Deficit Disorder are flooded by internal or external overstimulation and become hyperactive, distractible, or shut down as ways of coping with such stimulation. They would almost never be labeled self-centered or narcissistic. They display no psychotic symptoms and almost never express any suicidal ideation, other than as a transient cry for help.

Bipolar children are usually less overstimulated by external circumstances, though they do display distractibility and are very pleasure-seeking. They are much more self-absorbed, preoccupied with their own internal need fulfilment, and are more narcissistic. They are capable of being, at times, morbidly preoccupied with their own inner thoughts, sometimes culminating into either suicidal ideation or even intention. Their affect or feeling tone may be inappropriate as a result of an incongruency of mood and cognition. In extreme cases, misperception of reality can reach psychotic proportions.

Children with attachment difficulty have self-centered primitive process distortions. Their trauma history influences all of their perceptions and they are motivated to be on the eternal lookout to keep themselves safe.

Peer Relationships

There can be widely varying styles of relationships in both Attention Deficit and bipolar children. Attentional Deficit Hyperactivity Disordered children can range from being very "cool" and nondemanding to very "dorky," displaying innocent and immature styles of relating. They may have a tendency to make friends easily but may have difficulty maintaining them.

Bipolar children, depending upon their mood, can range from exceedingly charismatic and enthusiastic, to being capable of severe conflicts because of their highly calculating, controlling, and manipulative styles of relating. At other times, they may be isolative or withdrawn. They consistently have high irritability levels so that conflicts with other children are common.

Children with attachment difficulties have consistently poor peer relationships. They will target others to get angry at them. Their consistent nonintimacy and control issues limit any quality of friendships.

Sleep Patterns

Being overstimulated and physically wound-up, children with Attention Deficit Disorder display occasional difficulty getting to sleep. Once most ADHD children get to sleep, they commonly "sleep like a rock," though they may be squirmy or fidgety in their sleep. Recurrent nightmares are uncommon.

Many children and adolescents with Bipolar Disorder are "night people," commonly preferring to stay up late. They may even display a greater clarity of thought into late evening. They have difficulty emotionally and physically relaxing, "winding down," and going to sleep. While their mind may race, they have difficulty explaining or expressing this phenomenon. Some bipolar children require very little sleep and others may have excessive sleep patterns to the point of hypersomnolence. Recurrent nightmares are common. They are capable of having vivid dreams, at times even gory nightmares.

Hypervigilance creates very light sleepers in attachment-disordered children. They tend to require less sleep and commonly rise early in the morning.

Motivation

Some initial ambitiousness is common in children with Attention Deficit Disorder, unless their self-esteem is low or they have performance anxiety. Their difficulty is with task completion. Self-directed behavior is difficult. They are less individually resourceful and more adult dependent. Most children with Attention Deficit Disorder want to please adults.

Children with Bipolar Disorder often have exaggerated expectations of their own resourcefulness and are insistent on self-directed behavior. They may have highly variable energy outputs and can go from extremes of motivated, self-directed, goal-directed behavior to minimal motivation, interest, and enthusiasm, at times even with those activities they had previously demonstrated interest. Many bipolar children are capable of showing giftedness, with both creativity and verbal articulativeness being common.

Consistently poor initiative, limited industriousness, and intentional inefficiency are common characteristics of children with attachment difficulties. They are motivated for the short-term only.

Learning Characteristics

Multiple studies demonstrate up to a 50% correlation of Attention Deficit Disorder with specific learning disabilities, most commonly auditory perceptional difficulties and fine motor incoordination. Handwriting is often difficult. These are the classic "right-brained" children. Other learning difficulties are not unusual for individual children with Attention Deficit Disorder.

Children with Bipolar Disorder are nonsequential, nonlinear learners. They are at times able to grasp large amounts of information in very brief periods of time, making leaps of knowledge. At other times, they are unable to grasp any detail and their motivation to learn is low. Bipolar children will sometimes use their new knowledge in shrewd or conniving ways.

Brain maturational delays secondary to maternal alcohol or drug effects or early life abuse and neglect can create multiple and diverse learning problems in attachment-disordered children.

Anxiety

Children with Attention Deficit Disorder seldom manifest much anxiety, unless it is specifically performance related. Generalized anxiety states are very uncommon in these children.

Children with Bipolar Disorder are emotionally wired and have high potentials for anxiety, tension, and agitation. Specific fears, phobias, and obsessions are well known and documented with these children. Their anxieties can be so severe that they are capable of dissociation. Being internally preoccupied, they are prone to have physical or somatic complaints. Needle phobias are common.

Children with attachment difficulties have very poor conscious recognition, awareness, or admission of anxiety or fears. Their hypervigilance is a symptom however, of substantial anxiety. They appear invulnerable. Unconsciously felt anxiety will commonly contribute to dissociation.

Sexuality

Attention Deficit Disordered children are emotionally immature and sexually naive. They are rarely the first children in elementary, middle, or junior high school to become either sexually interested or involved. Most adolescents with attentional problems have enough reality orientation and conscience development to not engage in sexual activity during times of potential ovulation or may do so with the use of birth control methods.

Children with Bipolar Disorder, on the other hand, are commonly sexually hyperaware and demonstrate either precociousness or pseudomaturity. They are highly sexually curious and are among the first within different age ranges to demonstrate sexual interest and activity. There is a high incidence of adopted children in general with Bipolar Disorder. One of the reasons is the rate of very high adolescent sexuality among bipolar boys and girls. Being intensely mood driven, they will disregard risk factors of ovulation and have high rates of teenage pregnancies as a result.

Children and adolescents with attachment disorders use sex as a means of power, control, or infliction of pain, including sadism.

Alcohol and Substance Abuse

Alcohol and substance abuse is very common in adolescents with Attention Deficit Disorder. They abuse out of a sense of experimentation, attempting to be like others and to impress their friends. These substances also serve as coping mechanisms for low self-esteem.

Adolescence with Bipolar Disorder have exceptionally high rates of alcohol and substance abuse, many studies showing up to a 60% correlation. On one hand they may use these substances in an attempt to "treat" either hypomanic or depressive mood states. In reality, the facts are counterintuitive. It would be intuitive for hypomanic adolescents to seek out alcohol and other depressants to calm themselves down and use stimulants when depressed. I have seen many cases of exactly the opposite being true. It is my experience that hypomanic adolescents will actually seek out stimulant medications, as opposed to depressants, in order to enhance and augment an already excessively high energy state.

This extreme tendency for bipolar adolescents to abuse substances is an issue that cannot be overemphasized in terms of its ramifications, particularly when dealing with children of abuse. Given their high rates of sexual activity and high rates of alcohol and substance abuse, they are extremely prone to becoming pregnant, having children, and abusing or neglecting those children enough to create attachment symptoms. Nothing is worse than these children having children, driven by alcohol, substances, and extremes of mood. The potential for abusive or neglectful circumstances is extremely high. This group of drug-abusing bipolar adolescents or young adults is the single greatest population in the highest risk of maltreating their children.

A number of experts on attachment disorders were consulted about adolescents' potential to abuse alcohol or other substances. The consensus was that abuse seems to be sporadic and less common than adolescents with either ADHD or Bipolar Disorder. They do not appear to be able to lose much control. More research is needed.

Parenting Techniques

Parents with children or adolescents with Attention Deficit Disorder achieve positive effects with a skillful balance of love, support, patience, and a sense of humor along with the capacity to be directive, structuring, and deal with positive and adverse consequences in a timely manner. Parents achieve beneficial results by emphasizing quality, not quantity.

It is very difficult for parents of children or adolescents with Bipolar Disorder to have a consistent, positive effect on their child over the long-term. Different creative techniques will work for brief periods until the bipolar child, invested in being in control, finds some new way to defy or obstruct. These characteristics are very frustrating for almost all parents. Only when a correct diagnosis and appropriate and effective medical treatment occurs do these children become more compliant and agreeable.

Parenting children with attachment difficulties can range from gratifying and heartwarming to (with the more severely attachment-disordered children) enormously confusing, frustrating, and even hostility-provoking. Parents frequently blame themselves and/or are judged within their communities to be inadequate parents. Many of these latter children are workable, however, depending upon multiple factors. The child's age, duration, extent, and intensity of abuse and/or neglect, and the child's innate temperament and community support are all important factors. Working with a knowledgeable and effective psychotherapist who has familiarity with attachment disorders is essential.

Optimal Environment

Children with Attention Deficit Disorder have enhanced functioning in a relatively low-stress, low-stimulation, encouraging environment. Being approval seeking, they respond well to affection. It is also important to identify learning style variations, including learning disabilities, as well as identify and treat psychological factors, such as low self-esteem and performance anxiety.

Bipolar children have a more unpredictable course and are less receptive to encouragement, support and affection. Even though they may cognitively desire approval, their actions remain mood driven and intense. They require, on one hand, clarity and assertion of limits and, on the other hand, flexibility in reaching some negotiation. They need to be contracted with because they need to make the decision to comply or defy.

Children with attachment difficulties require a challenging balance of support, security, stability, consistency, predictability, and dependability. These environments enable the child's self-esteem to rise and trust in others to develop. All of this requires a certain unambiguity of expectations, a sense of humor, nurturance, and love.

Psychopharmacology

Medications helpful for children with Attention Deficit Disorder include the classic use of stimulant medications. Methylphenidate (Ritalin) is the classic and still most widely used medication for the treatment of ADHD. The inclusion of Adderall within the past few years as well as the traditional use of dextroamphetamine (Dexedrine) is very helpful. Pemoline (Cylert) is a longer-acting stimulant medication. All of these medications are useful but have occasional side-effects such as appetite suppression and enhancement of sleep disturbance. Inclusion of bupropion (Wellbutrin) is a very useful addition, particularly for those children with combined mood, motivational, and attentional symptoms. Inclusion of either clonidine (Catapres) or guanfacine (Tenex) may have ancillary effects on hyperactivity, anxiety, or aggressive symptoms.

Medications helpful for Bipolar children and adolescents include the use of mood-stabilizing medications. These are lithium (Eskalith, Lithobid), valproate (Depakote), verapamil (Calan, Verelan), carbamazepine (Carbatrol, Tegretol), and gabapentin (Neurontin). Medications that are most helpful for moderating severe opposition and rage reactions include olanzapine (Zyprexa), quetiapine (Seroquel), and risperidone (Risperdal). At the time of the writing of this chapter, ziprasidone (Zeldox) was on a track toward acceptance by the Federal Drug Administration. Antidepressants, especially bupropion (Wellbutrin), may be helpful for mood and motivational enhancement. While serotonergic antidepressants have been helpful for this, they are also more prone, in my experience, than bupropion to destabilize mood.

Children with attachment disordered symptoms may also be helped by clonidine and guanfacine. These medications can reduce the hypervigilance. If there is no bipolar component, antidepressants may be helpful at moderating hypervigilance, hyperarousal, and other associated posttraumatic symptoms.

Prognosis

Children with Attention Deficit Disorder have a good to excellent prognosis with appropriate medical treatment and other ancillary therapies and educational accommodations.

Children with Bipolar Disorder have prognoses that range from fair to good. They are prone to periods of regression, even with appropriate and otherwise effective treatment.

The prognosis for children with attachment difficulties is highly variable and is dependent on multiple factors. Among these are the recognition of a comorbid mood disorder, degrees of intensity and frequency of abuse or neglect, the age of relinquishment, innate temperament, and quality of treatment.

EFFECTIVE MEDICAL TREATMENT PLANS: PSYCHOPHARMACOLOGY OF DISRUPTIVE BEHAVIORAL DISORDERS ASSOCIATED WITH EARLY-LIFE MALTREATMENT

An effective medical treatment plan consists of the use of medication or commonly the combination of medications. These children present in variable and complex ways and the use of medications, and their favorable side-effects, often result in moderate to substantial improvement in their symptom complex. Obviously, an effective medical treatment plan is dependent upon an accurate medical diagnosis. As previously elaborated, medical treatment plans that are based inaccurately on diagnoses of Attention Deficit Disorder, Dysthymic Disorder, Major Depression, or Post-traumatic Stress Disorder in disruptive behaviorally disordered children are likely to not work well over extended periods. As such, the following is a brief outline of different types of medications potentially useful for this group of children.

While there is a lack of published double-blind, placebo-controlled studies of medications for children and adolescents with Bipolar Disorder, most physicians extrapolate from data on the successful pharmacological treatment of adults with Bipolar Disorder. Some of these medications require laboratory testing before and during treatment and all require careful management and monitoring for dose regulation and side-effect potential. The guideline that I always use in determining dosage of medication is "the least amount that does the most good." Careful monitoring and feedback from caregivers will help. In my experience, the dosages needed of various medication is on a skewed bell shaped curve, with some children responding to low dosage, others to moderate dosage, and still others to high dosages. My experience is that these children as a group respond to medication dosages within the higher dosage range. Their early age of onset of severe symptoms is indicative of severe biochemical imbalances. These children uncommonly manifest mild symptoms. They tend to be severely disturbed. There is scientific data to suggest the earlier and more severe the onset of symptoms, the poorer the overall prognosis. These children are also commonly in generally good physical health, and their metabolic healing capacities are significant. They have "super livers" and "super kidneys," metabolizing and clearing these medications far faster than adults. As such, they require relative and literal higher equivalent doses than adults. It is my experience that the average child actually has fewer side-effects than an adult would given the same medication.

Children often benefit more quickly from medication trials than adults do from trials of the same medications. Depending on the medication prescribed, it is common to get a clarity of direction of positive or negative outcomes within a few weeks. A determining factor is the presence or absence of side-effects as the child starts on medication. Obviously, it is prudent to adjust dosages upward more slowly if side-effects are troublesome or more expeditiously if side-effects are absent to easily tolerable. Most trials of medication do not need to be beyond 4 to 6 weeks. Conversely, I have had a few children respond optimally following 2- to 3-month trial periods.

Combinations of various medications, such as exemplified in Jim's case history, are often necessary. Combinations of mood-stabilizers as well as mood-stabilizers with either atypical antipsychotic medication or antidepressants or even other medications are frequent. Especially in treating the child with comorbid attachment symptoms and Bipolar Disorder, more than just childhood Bipolar Disorder alone, it is frequently necessary to combine medications with choices dependent upon symptoms presentation. These combinations are more capable of treating diverse aspects, such as labile mood, dysphoria, and overt opposition and rage. The combination of medications also often enables the clinician to use lower doses of two medications than the clinician would with one or the other. Often combinations are complimentary; sometimes they are not. Trial and experience is necessary.

The following is an overview of available medications. It is not intended as an exhaustive treatise but to guide the reader regarding available options. No specific endorsement is intended or implied toward any specific name-brand medications.

Mood-Stabilizing Medications

These medications, as a whole, are the foundation of the treatment for Bipolar Disorder. Some of these medications have been available for decades, still others only recently. Lithium Carbonate (Eskalith, Lithobid) until recently was historically most commonly used to stabilize mood in children. This medication is usually not my first choice because of some troublesome ongoing side-effects, such as thirst and weight gain. Lithium, in contrast to all of the other mood-stabilizing medications, also has the greatest long-term potential to suppress thyroid function. However, I have had large numbers of children and adolescents who have optimally responded to lithium, compared to the other available mood-stabilizers. Because of rapid kidney clearances of this medication, it is helpful for children and younger adolescents to take this medication, even the sustained-release forms, up to three times daily to maintain adequate serum (blood) concentrations.

Valproate (Depakote) has recently overtaken lithium as the first-line prescribed mood-stabilizer. An antiepileptic medication, it has been shown to have highly effective mood-stabilizing properties. Its effectiveness is equivalent to lithium and, in general, it has a more favorable side-effect profile. There is evidence that it may be more effective than lithium for the "mixed type" of Bipolar Disorder, which is how the vast majority of children present. It commonly does not create a dulled effect, with improvements in cognitive brightness and clarity. Calming effects are well known. In my experience, though caution should be used in prescribing to very young children, concern for severe side-effects, especially liver and blood problems, has been overrated. Weight gain can be a troublesome, though tolerable, side-effect.

Verapamil (Calan, Verelan) is a highly effective mood-stabilizing medication that is substantially underprescribed in children, adolescents and adults. A calcium channel blocker which is principally used for blood pressure regulation, this medication is nonetheless effective at stabilizing mood. Because of children's malleable and flexible cardiovascular systems, they are seldom prone to side-effects such as light-headedness or dizziness that affect adults with high blood pressure taking this medication. It is available in immediate- and extended-release forms. The side-effects, other than transient constipation, tend to be highly tolerable.

Carbamazepine (Carbatrol, Tegretol) has been known to be a valid and helpful mood-stabilizing medication. This is a medication in which the trademark sustained-release forms are superior, in my opinion, to the generic immediate-release forms. It has potential advantages over lithium and valproate in that it does not cause weight gain. Short-term memory loss and word retrieval difficulties are well known possible side-effects. Rare side-effects can be dangerous with this medication, and require careful monitoring.

Gabapentin (Neurontin) has excellent mood-stabilizing properties and is substantially underutilized and underprescribed in my opinion. It appears to be a highly safe mood-stabilizer, its side-effect profile being generally very low. In contrast to some of the other mood-stabilizers, which can have a mentally dulling effect, gabapentin can even be mentally clarifying and physically energizing in addition to having calming properties. Because of its mildness, it commonly works best if added to other medications.

Atypical Antipsychotic Medications

As a broad group, these medications have been revolutionary and "miraculous," especially in their treatment of the symptom continuum of severe irritability, opposition, defiance, temper outbursts, rage reactions, and verbal and physical aggressivity, including assaultiveness. No other group of medications is as effective, in my experience, for this continuum of symptoms than these group of medications. Indeed, within the past few years, when the child's presentation is one of severe rage and the family is in crisis with potentials of disruption, it has occasionally been my practice to prescribe this group of medications initially, often with highly effective results. These medications have not only the potential to substantially modify the opposition and rage component, but they have often resulted in more clear-headed, motivated children with a genuine desire to want to please.

This class of medications is potent and associated with rare, severe side-effect potentials and other disorders. To moderate the potential for these side-effects, I recommend all patients on this group of medications also be on Vitamin E. Because one is commonly looking at severe disruptive behavioral disorders, it is my experience that these medications, when prescribed by physicians, are commonly prescribed at inadequate doses. I have had many cases where a previous treating physician prescribed the right medication but in such small doses that they never achieved a desirable outcome. Obviously, it is always important with all medications, not just medications of this type, to use the least amount that has the most desirable therapeutic effect.

Olanzapine (Zyprexa) has been extremely helpful at substantially modifying oppositional and rage tendencies while producing mentally clarifying and mood-enhancement effects in both children and adolescents. It is equally effective in its therapeutic benefits as the other atypical antipsychotic medications. Its major drawback is the potential for substantial weight gain as a common side-effect.

Quetiapine (Seroquel) is the newest member of this class of medications. It has the distinct advantage of having the lowest side-effect profile of any of the medications available within this group. The disadvantage of using this medication is that there can be a broad range of effective dosages with tendencies for therapeutic effects to rise as dose is increased, to a point. In other words, it is my experience that people need well above "average" recommended doses with this medication to achieve desirable benefits.

I have had occasion to use Risperdone (Risperdal) with over 200 bipolar children and adolescents. I have used it because of its long-term availability, effectiveness and generally tolerable side-effect profile. Weight gain is a serious potential side-effect associated with this medication but statistically less so than with olanzapine. There have been a few early-adolescent girls who have been prone to sexual side-effects with this medication.

At the time of the writing of this chapter. Ziprasidone (Zeldox) was in expanded trials in the pharmaceutical company's attempt to obtain approval from the Federal Drug Administration. If approval is granted, this medication will be added to our medical armamentarium. While an effective atypical antipsychotic medication, it is said to also produce positive elevations of mood. Ziprasidone also has less potential for weight gain than other members of this class.

Antidepressant Medications

Antidepressant medications are occasionally needed in the treatment of disruptive behavioral disorders. In contrast to some physicians who find them helpful at moderating opposition and rage, I find that, more often than not, they have a destabilizing effect on mood. I have been told on numerous occasions how antidepressant medications, when used alone, have resulted in a decrease in stability in these children. I do use antidepressant medications, however, in those children and adolescents where there are strong dysphoric symptoms. These would include high levels of unpleasability, unsatisfiability, and depressed mood, occasionally culminating in suicidal ideation or intention. They are also helpful when the child is doing well, but maintains low levels of interest, motivation, enthusiasm, or sustained low mood. They are to be added to mood-stabilizing and/or antipsychotic medication, not used instead of these medications.

Bupropion (Wellbutrin) is very helpful at improving mood, motivation and overall interest and enthusiasm levels. Its overall activating effects can, at times, help counteract dulling or sedating effects of mood-stabilizing or atypical antipsychotic medications. Its side-effect profile is low though caution should be used in prescribing this medication to children with a seizure history or strong compulsive components, especially eating disorders.

There are currently five medications within the class of Selective Serotonin Reuptake Inhibitors (SSRI). They have very similar mechanisms of action and generally similar side-effect profiles, although there are some specific minor differences. I have used these medications sparingly in the treatment of child and adolescent Bipolar Disorder, Disinhibited Type, associated with early life abuse and neglect. These medications are helpful in the group of Bipolar Disordered children and adolescents with very obsessive-compulsive components to their symptom profile. These children have ongoing, unwanted obsessional thoughts and carry on ritualistic, compulsive deeds. Side-effect profiles with all of these medications are favorable, particularly in children and younger adolescents. When used, these medications are best used in small doses, as higher doses can actually contribute to mood instability.

These medications have well-known names, such as fluoxetine (Prozac), paroxetine (Paxil), and sertraline (Zoloft). Fluvoxamine (Luvox) is not specifically approved as an antidepressant but is approved for Obsessive Compulsive Disorder. It is probably equally effective as an antidepressant. The recent availability of citalopram (Celexa), long available in Europe, gives even broader choice in the selection of medication.

Venlafaxine (Effexor XR Capsules) in the extended-release form has similar serotonergic activity as the SSRI medications in low doses. In higher doses, it also stimulates noradrenaline. It is a potent medication with a favorable side-effect profile.

Nefazodone (Serzone) and mirtazapine (Remeron) are two "novel" antidepressant medications. I have had excellent outcomes in a few children with the combination of nefazodone and risperdone. I have yet to use mirtazapine in these children.

Tricyclic Antidepressants are the classic antidepressants that have been in clinical use for decades. The two main medications in this class are imipramine (Tofranil) and nortriptyline (Pamelor). Despite their traditional multiple uses for depression, ADHD, panic disorder, conduct disorder, and enuresis, I have been reluctant to prescribe these medications for several reasons. They have multiple side-effects, including dry mouth, blurred vision, and weight gain. There are known potentials for cardiac toxicity. These medications are potentially lethal in overdose. Last, it has been my experience that these medications can result in switching from depression to mania and contribute to rapid cycling.

α-Adrenergic Agonists

This group of medications has two principle drugs: clonidine (Catapres) and guanfacine (Tenex). These medications, both blood pressure medications, have substantial benefits of moderating any or all "arousal" symptoms, such as hyperactivity, hypervigilance, anxiety, agitation, or aggression.

Clonidine, the historical standard treatment, is helpful at moderating all of the above symptoms. Clonidine's most troublesome side-effect is sedation and there are frequent limits to its effectiveness because of this side-effect, which may or may not wear off over several weeks. Infrequently, dizziness or light-headedness are also troublesome.

Within the past few years I have tended to prefer guanfacine over clonidine. It has fewer and less significant side-effects, particularly sedation and light-headedness. Because it is less sedating, commonly one can reach more therapeutic dose levels with guanfacine than with clonidine.

Stimulants

This group of four medications are well known for their benefits in children with Attention Deficit Disorder. However, as previously stated within the text of this chapter, I have found them of benefit in about 5% of children with Bipolar Disorder with comorbid Reactive Attachment Disorder. Being stimulants, they are capable of accentuating and exacerbating already arousable children to even higher levels of arouseability and excitation, "kindling" hypomanic symptoms. I have had at least 100 mothers of children with undiagnosed Bipolar Disorder whose physicians mistook these children as having Attention Deficit Disorder describe severe accentuation of symptoms with the addition of stimulant medications. Many of these children never returned to their previous baseline, even upon discontinuation of their stimulant medications. These medications may also contribute, both initially and long-term, to side-effects of appetite suppression, weight loss, irritability, emotional oversensitivity, and insomnia. Some bipolar children placed on stimulant medication develop an extremely low mood.

Methylphenidate (Ritalin) is the classic stimulant and is more commonly prescribed than any of the other stimulant medications. It has a very short half-life, especially in its immediate-release form, but even in its sustained-release form does not tend to last long in the body. In my experience, methylphenidate has the greatest degree of tolerance associated with its use. In other words, even when methylphenidate is effective, it is common for people to have to continue to raise the dose of this medication periodically to maintain desirable effects.

Dextroamphetamine (Dexedrine) is a very worthwhile and, in my opinion, underprescribed medication for Attention Deficit Disorder. However, it is not often used in treating childhood Bipolar Disorder.

While Pemoline (Cylert); is milder than the other stimulant medications, the same cautions apply with the use of this medication in childhood Bipolar Disorder.

Adderall is a relatively recent addition to the pharmacological treatment of ADHD. It consists of two different salts of both dextroamphetamine and methylamphetamine. In my experience, this combination produces a "mellow" effect in the treatment of attentional disorders. It has a gradual onset of effects, a sustained optimal period of effectiveness, and a gradual release from the body. It is becoming a popularly prescribed medication for ADHD. While a worthwhile medication in itself, it is not a standard part of a medical treatment regimen in the treatment of a mixed type of Bipolar Disorder.

β-Adrenergic Blocking Agents

Propranolol (Inderal) is the only common medication of this type used in children and adolescents. In adults, it has been prescribed for decades for high blood pressure, cardiac arrhythmias, tremor, and other problems. This medication may be of some benefit in the management of aggressive and assaultive behaviors in general, although I have not found it to be particularly beneficial for those symptoms in bipolar children. I have found this medication to be of benefit in the management of infrequent tremor as a result of different combinations of prescribed medications. For example, the use of lithium, valproate, or carbamazepine in combination with each other or in combination with risperidone, olanzapine, or quetiapine may occasionally result in hand tremor as a troublesome side-effect. This tremor is almost never a resting tremor. Instead, the tremor produced is an "intention tremor." It is most common in voluntary motion, such as handwriting, feeding oneself, or drinking with a glass. During these times, usually low doses of propranolol have had an outstanding potential to moderate, if not eliminate, these regrettable tremors.

Antianxiety and Sedative Medications

Benzodiazepines were the most commonly prescribed—and potentially abused—drug in the United States from the late 1960s to 1980. They would be well known by at least their trademark names, if not their generic names. This group of antianxiety medications include alprazolam (Xanax), chlordiazepoxide (Librium), clonazepam (Klonopin), clorazepate (Tranxene), diazepam (Valium), and lorazepam (Ativan). The sedative part of this family includes flurazepam (Dalmane), quazepam (Doral), temazepam (Restoril), and triazolam (Halcion). While predictably effective in a very high percentage of patients, these medications possess several drawbacks both in the short and long term. These include sedation, drowsiness, and confusional states. As CNS depressants, up to 30% of people taking any of these medications in an ongoing manner could become tolerant or physically habituated. Lastly, as depressants, they can worsen disinhibition. As such, they have little or no use in this patient population.

Buspirone hydrochloride (BuSpar) is an antianxiety medication that is not a CNS depressant. As such, it produces no potential tolerance or habituation. It is less sedating than the benzodiazepines and has low abuse potential. Its overall benefits are its general mild effects, which can cut both ways. Being mild, it may not have enough potency to warrant its use. Several children have benefitted, however, from the addition of this medication to their medication regimes.

"Classic" Neuroleptic Antipsychotic Medications

I am including this group of medications in an attempt to be comprehensive. I have used this group of medications very infrequently since the introduction of the atypical antipsychotic medications several years ago. While they are highly effective at eliminating psychotic symptoms of delusions, hallucinations, and even aggression, their potential for a multitude of neurological or other side-effects is well known.

Tardive dyskinesia is a permanent neurological disorder associated with the long-term use of the high-potency antipsychotic drugs, such as haloperidol (Haldol), trifluoperazine (Stelazine), perphenazine (Trilafon), thiothixene (Navane), loxapine (Loxitane), fluphenazine (Prolixin), and pimozide (Orap). Since Bipolar Disorder is a lifelong, chronic condition, the use of medications that potentially enhance the potential for severe permanent neurological disorders is to be avoided, if possible. In consulting with various attachment centers, I have been fortunate to consult with children from all over the United States. There are clusters of states within the United States where physicians are still prescribing these medications to large numbers of children. In my opinion, this medical practice is ill advised. There are much more effective and less troublesome medications.

There are low-potency antipsychotic medications, such as chlorpromazine (Thorazine) and thioridazine (Mellaril) that are far less likely to create either extrapyramidal side-effects or influence the development of tardive dyskinesia. They are also more likely to produce sedation, low blood pressure, dry mouth, and provoke sexual dysfunction. While of limited pharmacological benefit, considering their side-effect potential, their primary benefit is low financial cost. They may also be of some benefit in that group of children where almost all other treatment options have been found to be ineffective or in those children who are in crisis circumstances.

Characteristics of Attention Deficit Disorder, Bipolar I Disorder and Reactive Attachment Disorder

Symptoms	Attention Deficit Disorder, with or without hyperactivity	Bipolar I Disorder, mixed type (rapid cycling)	Reactive Attachment Disorder, disinhibited type
Age of onset	Infancy—toddler, 6, 13	2–3, 6, 13–25	Birth to 3
Family history	ADHD, academic difficulties, (based on task incompletion) alcohol and substance abuse.	Any mood disorder (depression or bipolar), academic difficulties, based on motivation problems or opposition or defiance, alcohol and substance abuse, adoption, ADHD.	Abuse and neglect, severe emotional and behavioral disorders, alcohol and substance abuse, abuse and neglect in parent's own early life.
Lifetime prevalence	Approximately 3–6% of general population	1–3% of general population	Uncommon to common.
Etiology	Genetic, neurochemical, fetal development, brain traumas, nutritional deficiencies, exacerbated by stress.	Genetic, exacerbated by stress and hormones.	Psychophysiologic secondary to neglect, abuse, mistreatment, abandonment.
Duration	Chronic and unremittingly continuous, tending toward improvement.	May or may not show clear emotional and behavioral episodes and cyclicity; worsens over years with increased severity of symptoms.	Dependent on life circumstances, age of relinquishment, including innate temperament and treatment. Worsens over years without treatment to develop antisocial character disorders.
Attention span	Short, leading to lack of productivity and task performance and completion.	Entirely dependent on interest and motivation, distractibility common.	Hyperarousal influences hypervigilance, distractibility and shortened periods of focus. Shortens with stress.
Impulsivity	Secondary to inattention or obliviousness, regret and remorse.	"Driven," "Irresistible," grandiosity, thrill-seeking, counterphobia, little regret or remorse. Pressured speech.	Poor cause and effect. No remorse. Can range from overreactive to highly controlled, self protective.
Hyperactivity	50% are hyperactive. Disorganized, fidgety, jittery.	Wide ranges, with hyperactivity common in children.	Common
Self-esteem	Low, rooted in ongoing performance difficulties.	Low because of inherent unpredictability of mood. Grandiose or expansive mood could mask low esteem.	Low, rooted in abandonment, feel worthless and unlovable, masked by anger or indifference.

Mood	Usually friendly in a genuine manner. Some irritability.	Unpredictable, oversensitive, expansive, grandiose, irritable, hard to please or satisfy.	Superficially charming, phoney, distrusting, emotionally distant, nonintimate.
Control issues	Desire to seek approval—get into trouble by inability to complete tasks.	Intermittently desire to please but tend to push limits and relish power struggles. Expert hasslers, persuasive.	Controlled and controlling, only for self-gain, underhanded, sneaky and covert.
Opposition/Defiance	Demonstrate argumentativeness but will relent with show of authority, redirectable. Short attention span allows them to "let go" easily.	Usually overtly and prominently defiant, at times passive aggressive, often not relenting to authority. Tend to insist on getting own way.	Conning and cunning. Covertly defiant, passive aggressive.
Blaming	Self-protective mechanism to avoid immediate adverse consequences.	Grandiosity contributes to disbelief/denial they caused something to go wrong.	Rejecting of responsibility. Victim position.
Lying	Avoid immediate adverse consequences.	Enjoys "getting away with it."	"Crazy lying," stuck in perceptual self-centered "primary process" distortions in order to attempt to gain control.
Fire setting	Play with matches out of curiosity, nonmalicious.	Intrigued with matches/fire setting and can have malicious intent.	Revenge motivated, malicious. Danger seeking secondary to despair.
Anger, irritability, temper, and rage	Situational, in response to over-stimulation, poor frustration tolerance and need for immediate gratification. Rage reaction is usually short-lived.	Secondary to limit-setting or attempts to control their excessive behavior, rage can last for extended periods of time, at other times may be explosive and over quickly. Overt, aggressive, and assaultive.	Chronic, revenge "get even" oriented. Eternal "victim" position, with rationalizations for destructive retaliation. Hurtful to innocent others and pets.
Entitlement (deserving of special benefits)	Overwhelming need for immediate gratification (not a prominent symptom).	Expansive and grandiose mood creates belief they deserve special treatment.	Compensation for abandonment and deprivation (not a prominent symptom).
Conscience development	Capable of demonstrating remorse when things calm down. Close to developmental age.	Limited conscience development, dependent on mood and parenting ability.	Very "street smart," good survival skills, con artists, calculating, devious.

(*continues*)

Characteristics of Attention Deficit Disorder, Bipolar I Disorder and Reactive Attachment Disorder *(continued)*

Symptoms	Attention Deficit Disorder, with or without hyperactivity	Bipolar I Disorder, mixed type (rapid cycling)	Reactive Attachment Disorder, disinhibited type
Sensitivity	Oblivious to detailed circumstances they are in and inappropriateness shows as result. Do get "big picture."	Acutely aware of circumstances and are "hot reactors."	Hypervigilent, compensating for past helplessness. Resistant and insensitive, rarely ill. Limited emotional repertoire.
Perception	Flooded by sensory over-stimulation, distractible, hyperactive or shut down.	Self-absorbed, preoccupied with internal need fulfillment, appears narcissistic. Dissociation possible. Inappropriate affect.	Self-centered primary process primitive distortions. Dissociation possible.
Peer relationships	Makes friends easily but not able to keep them. Immature.	Can be charismatic or depressed, depending on mood—conflicts are the rule due to controlling nature.	Very poor. Secondary to lack of intimacy and control issues. Target others to get angry. No long term friends.
Sleep patterns	Occasional trouble getting to sleep. Over-stimulated and physically wound up—once asleep "sleep like a rock." Fidget even in sleep. Nightmares uncommon.	Inability to relax, wind down because of racing thoughts—nightmares common.	Hypervigilance creates light sleepers. Tends to need little sleep, arise early in A.M.
Motivation	Less resourceful—more adult dependent. Okay starters, poor finishers.	Grandiose—believe they are resourceful, gifted, creative. Self-directed, highly variable energy and enthusiasm.	Consistently poor initiative, limited industriousness, intentional inefficiency. Motivation for short term only.
Learning characteristics	Most common auditory perceptual difficulties and fine motor incoordination. "Right brained."	Non-sequential, non-linear learners. Verbally articulate, used in shrewd and manipulative ways.	Brain maturational delays secondary to maternal drug/alcohol effects, early life abuse/neglect can create diverse learning problems.

Anxiety	Uncommon, unless performance-related.	Emotionally wired. High potentials for anxiety, fears, and phobias. Somatic symptoms common, needle phobic. Dissociation.	Appear invulnerable. Poor recognition, awareness or admission of fears. Dissociation.
Sexuality	Emotionally immature and sexually naive.	Sexual hyperawareness, pseudo-maturity, high interest and activity level.	Uses sex as means of power, control or infliction of pain, sadistic.
Alcohol and substance abuse	Strong tendencies, out of coping mechanisms for low self-esteem.	Very strong tendencies in attempt to enhance or reduce hypomanic/dysphoric moods.	Sporadic/uncommon; not likely to lose too much control. We need more knowledge of correlation.
Parenting techniques	Support, encouragement, redirection.	Nothing works long term until correctly diagnosed and medically treated.	Understanding child's vulnerabilities and resistances aids in being workable.
Optimal environment	Low stimulation and stress, support and structure. Identify learning disability components or psychological factors.	Clear and assertive, balance of limits, encouragement, negotiation. Helpful if all members of treatment team work together.	Challenging balance of security, stability, clarity, and unambiguity of expectations, nurturance, encouragement, and love.
Psychopharmacology	Medications helpful include Ritalin, Dexedrine, Cylert, Adderall, Wellbutrin. Clonidine and Guanfacine may be useful as additive medications.	Medications helpful to stabilize mood include: Lithium, Valproate, Verapamil, Carbamazepine and Gabapentin. Medications helpful for opposition and rage reactions include: Olanzapine, Risperidone, Quetiapine, and Ziprasidone. Antidepressants in small doses helpful for mood and motivational enhancement.	Antidepressants, Clonidine, Guanfacine may help decrease hypervigilance. Medications do not decrease characterological traits.
Prognosis	Good to excellent with appropriate medical treatment, ancillary therapies, and educational accommodations.	Fair to good with times of regression even with appropriate treatment.	Highly variable, dependent upon recognition of comorbid mood disorders, degree of abuse/neglect, age of relinquishment, innate temperament, and treatment.

REFERENCES

American Psychiatric Association (1994). Diagnostic and statistical manual of mental disorders (4th ed.). Washington, DC: Author.

RECOMMENDED READINGS

Ainsworth, M. D. S., Bleher, M., Waters, E., & Wall, S. (1978). *Patterns of attachment.* Hillsdale, NJ: Erlbaum.

Alston, J., (1996). New findings in diagnosis: Correlation between Bipolar Disorder and Reactive Attachment Disorder. *Attachments Journal*, Winter. Also accessed through the internet at www.attachmentcenter.org.

Alston, J. (1997). *Characteristics of Attention Deficit Disorder, Bipolar I Disorder and Reactive Attachment Disorder.* Unpublished. Available upon request from author.

Alston, J. (1995). Defining the problem. In C. A. McKelvey (Ed.), *Give them roots, then let them fly: Understanding attachment therapy* (pp. 23–27). Evergreen, CO: Attachment Center Press.

American Academy of Child and Adolescent Psychiatry (October, 1997). Supplement—Practice parameters for the assessment and treatment of children and adolescents with Anxiety Disorders, 69S–84S.

American Academy of Child and Adolescent Psychiatry (October, 1997). Supplement—Practice parameters for the assessment and treatment of children, adolescents and adults with Attention Deficit/Hyperactivity Disorder, 85S–121S.

American Academy of Child and Adolescent Psychiatry (October, 1997). Supplement—Practice parameters for the assessment and treatment of children and adolescents with Bipolar Disorder, 157S–176S.

American Academy of Child and Adolescent Psychiatry (October, 1997). Supplement—Practice parameters for the assessment and treatment of children and adolescents with Conduct Disorder, 122S–139S.

American Academy of Child and Adolescent Psychiatry (October, 1998). Supplement—Practice parameters for the assessment and treatment of children and adolescents with Depressive Disorders, 63S–83S.

American Academy of Child and Adolescent Psychiatry (October, 1997). Supplement—Practice parameters for the assessment and treatment of children and adolescents with Posttraumatic Stress, 45S–26S.

American Academy of Child and Adolescent Psychiatry (October, 1997). Supplement—Practice parameters for the assessment and treatment of children and adolescents with Schizophrenia, 177S–193S.

BIOGRAPHY

John F. Alston

John F. Alston, M.D., is a child, adolescent, family, and adult psychiatrist in private practice in Evergreen, Colorado.

Dr. Alston earned his undergraduate degree from the University of Notre Dame and his medical degree from Tulane University School of Medicine. Dr. Alston completed a residency in general psychiatry from

Tulane University School of Medicine Department of Psychiatry and a fellowship in Child Psychiatry from the University of California School of Medicine in San Francisco, California. He is a member of numerous professional organizations, including the American Academy of Child and Adolescent Psychiatry, the Colorado Child and Adolescent Psychiatric Society, and ATTACh.

Since 1977, Dr. Alston has had a special interest in the evaluation and treatment of disruptive behavioral disorders associated with early life abuse or neglect. Within recent years, he has also had an interest in the differential characteristics of attentional disorders and mood disorders. Dr. Alston serves as a child psychiatric consultant to adoption, foster care, and attachment agencies.

Dr. Alston speaks nationally and internationally on these topics.

9

ATTACHMENT DISORDER AND THE ADOPTIVE FAMILY

TERRY M. LEVY
MICHAEL ORLANS

Evergreen Consultants in Human Behavior
Evergreen, Colorado 80437-2764

The vast majority of families seeking treatment at our clinic are adoptive families. The parents had adopted children with histories of severe and chronic maltreatment, multiple placements and caregivers, and compromised attachment. These children typically display oppositional, antisocial, and conduct-disordered behaviors and attitudes. The prevailing belief among clinicians and researchers has been that adopted children are at greater risk for developing an array of psychosocial disorders than nonadopted children (Brodzinsky & Schecter, 1990; Lindholm & Touliatos, 1980). For example, research has found that only 2% of children in the United States are adopted, yet approximately one-third of children placed in residential treatment centers are adopted. Additionally, adopted children have been found to be twice as likely to display psychosocial problems later in life and two to three times more likely to develop conduct disorders than their nonadopted peers (Jones, 1997).

The above-mentioned statistics are misleading. Recent research indicates that adoption itself does not necessarily result in severe psychosocial problems. Five key factors influence whether adoptees develop later problems. First, the nature of the *in utero* experience is a significant factor. Synchrony and bonding begin during pregnancy. Many observers and re-

Handbook of Attachment Interventions

searchers believe that the fetus feels loved and wanted when the mother and/or father instinctively use soft and soothing tones (Brazelton & Cramer, 1990). Women who want their babies and communicate love and acceptance have easier pregnancies and healthier infants than women with unwanted pregnancies (Verny & Kelly, 1981). Second, the child's age and developmental stage at the time of placement is crucial. Basically, the earlier a child is adopted the better the chances of secure attachment and normal psychosocial development. Third, is the history of abuse and/or neglect. Maltreatment results in emotional, social, and neurobiological damage that increases in severity as development unfolds. Fourth, the child's attachment history is significant. For example, the more moves, the higher the risk for compromised and disrupted attachment. Last, the quality of parenting after adoption does influence the child's adjustment. Parents who are emotionally and intellectually prepared and receive the necessary support and resources are better able to deal successfully with challenging adoption issues.

It is the specific population of adopted children with histories of maltreatment and severely compromised attachment that most often develop serious emotional and relational problems. These factors seriously affect children's ability to self-regulate, learn empathy and morality, accept limits and rules, and form close and trusting relationships. Many children adopted from the U.S. and abroad fall into this category and are coming to the attention of the mental health, social service, and criminal justice systems in increasing numbers (Levy & Orlans, 1998).

BONDING AND ATTACHMENT

Adopted children have significant connections to at least two family systems—one by biology, birth, and ancestry; the other by law, learning, and day-to-day parenting. All adopted children have prior bonds with their birth parents, but not all develop secure attachments with their adoptive parents (Watson, 1997).

It is important to distinguish between bonding and attachment, even though popular and professional language often refers to these two concepts interchangeably. *Bonding* is defined as a significant relationship that occurs without knowledge or conscious intent, not as a result of learning. Genetic bonding involves the sense of connection that exists as a result of common ancestry, inherited characteristics, and shared genetic background. Birth bonding involves the intensely emotional shared experience during pregnancy and birth between mother and infant (and father to some extent, as a member of the family system). *Attachment* is defined as the deep and enduring connection established between child and caregiver(s) in the first 3 years of life. It is a learned ability, the result of ongoing reciprocal interactions characterized by protection, need fulfillment, limits, love, and trust (Bowlby 1969, 1970; Levy & Orlans, 1998; Watson, 1997).

All adopted children have at least one significant loss—the loss of the genetic and birth bond with their biological parents and families. The "searching," either literal or emotional, done by most adoptees is driven by this loss (Brodzinsky *et al.*, 1992). Many adopted children have additional significant losses as well as maltreatment and trauma. Examples of multiple losses include: loss of early attachments to birth parents/families due to relinquishment or removal by protective services; loss of attachments to foster parents as they are moved through the system; loss due to prior failed adoptions, a result of the parents' inability to tolerate the ongoing acting-out of the adopted child.

THE *IN UTERO* EXPERIENCE

As noted above, bonding begins *in utero*. Research and observation have demonstrated the significance of the *in utero* experience. This is truly the dawn of attachment, the stage in which the baby-to-be and the parents begin the process of connecting. Pre- and perinatal psychologists, using modern clinical tools such as electronic fetal monitors and ultrasound, have proven that the unborn baby has well-developed senses and reacts to stimuli from mother and the external environment. Biochemical and emotional communication between parents (especially mother) and the fetus has significant impact on future development and health.

Every sensory system of the fetus is capable of functioning prior to birth; the unborn baby is keenly aware of his or her environment and reacts to changes in that environment (Samuel & Samuel, 1986). By the 5th month, the fetus can recognize the mother's voice and even shows a preference for different types of music—becoming more relaxed when listening to Vivaldi and more agitated when exposed to Beethoven or loud rock music (Verny & Kelly, 1981). The fetus decodes maternal emotions through a neurohormonal dialogue. When a pregnant woman becomes fearful or anxious stress hormones are broadcast throughout the entire mind/body system of both the mother and unborn child (Borysenko & Borysenko, 1994). Severe and chronic maternal stress is associated with prematurity, low birth weight, and infants who are hyperaroused and colicky (Brazelton & Cramer, 1990).

The unborn baby is particularly vulnerable to drugs, alcohol, and tobacco. *In utero* drug exposure often results in low birth weight, agitation, and various developmental impairments. Fetal Alcohol Syndrome and Fetal Alcohol Effect have long-term consequences on the child's ability to learn and develop (Besharov, 1994). Smoking injects the neonate with many toxic chemicals and dramatically decreases the oxygen supply. The fetus becomes agitated each time the mother even thinks about smoking a cigarette (Sontag, 1970). A recent study found that males whose mothers smoked during pregnancy had significantly higher rates of persistent violent criminal behavior as adults (Berman *et al.,* 1999).

ADOPTION AND THE TRAUMATIZED CHILD

A child's adaptation to the adoptive family is dependent, to a large extent, on the nature and quality of prior attachments and his or her reactions to separation and loss. Unresolved loss regarding prior attachments (healthy or unhealthy) inhibits the development of future attachment. Children with severely compromised and disrupted attachment are often placed with an adoptive family and, subsequently, do not form meaningful attachments to their new parents and siblings. They are incapable of trust and invite further abandonment and rejection. Fear and anxiety are hidden under angry, aggressive, and controlling behaviors.

Loyalty conflicts also prevent attachment in the adoptive family. These adoptees have not "come to terms" with the loss of birth and/or prior foster parents. Adoptive parents often feel threatened and confused by the child's reunification fantasies, the desire to maintain ties with prior attachment figures, and the tendency to place prior caregivers "on a pedestal." For example, a 10-year-old boy reveals during treatment his fantasy of reunification with birth mother "someday in heaven," despite the reality that he was repeatedly abused and abandoned by her. Children such as this commonly assume the role of "parental child"; they believe it is their responsibility to protect, nurture, and rehabilitate their birth parents. Their fantasized goal is to somehow "fix" the dysfunctional primary attachment figures, thereby developing a positive and complete sense of self: "I am now lovable and worthwhile."

Attachment-disordered adoptees have negative belief systems (negative working model); they perceive themselves as undeserving of love and are pessimistic regarding positive change. A vicious cycle ensues, whereby the child's negative belief system is inadvertently reinforced by parental reactions of anger and emotional distancing. A variety of powerful and confusing emotions, such as rage, sadness, fear, hopelessness, and shame are experienced by the children. They displace anger and aggression toward their adoptive parents rather than toward the perpetrators. Emotionally mature and stable adoptive parents are often confused about how to respond to this hostility and are frustrated by the constant rejection and attacks. Parents with unresolved emotional issues, however, are "triggered" by the child's relentless negativity, resulting in destructive responses and an ongoing negative parent–child relationship.

Moss (1997) reviewed the literature on the family qualities that increase the likelihood of successful adoption, with a focus on traumatized and attachment-disordered children. The key issues identified were:

1. Availability of a support system or support group.
2. Availability of pre- and postplacement services.
3. Possession and maintenance of good mental health, including tolerance of negative behaviors, a sense of humor, willingness to take time away from the children, and open communication in the marriage.
4. Parental role flexibility in rule-making and decision-making.
5. Family willingness to seek help in a crisis.
6. Parental knowledge of child development and attachment dynamics.
7. A positive perception of one's own parenting ability.

THE TRAUMATIZED ADOPTIVE FAMILY

Parents who adopt children with histories of severe maltreatment and attachment disorder are often intellectually and emotionally unprepared to deal with the extent of the child's problems and impact on the family system. Over time, chronic and severe stress levels associated with negative and destructive relationship patterns have created a climate of tension, despair, and hopelessness in the family. When the parents begin treatment with their child they are commonly angry, demoralized, and are experiencing family burnout. Figley defines family burnout as "the breakdown of the family members' collective commitment to each other and a refusal to work together in harmony as a function of some crisis or traumatic event or series of crisis or events that leave members emotionally exhausted and disillusioned" (Figley, 1998, p. 7).

It is common for parents and other family members (e.g., siblings) to experience secondary traumatic stress disorder (STSD), the result of the chronic tension and stress associated with the demands of living with and caring for the child with attachment disorder. This leads to emotional exhaustion, the breakdown of the family system, and family burnout (Figley, 1989, 1998). The psychosocial environment of the traumatized family is characterized by the following components and dynamics.

1. *Intimacy, affection, and emotional climate.* The emotional climate is characterized by anger, frustration, hopelessness, and despair. Children with histories of compromised attachment compulsively reenact negative patterns of relating. For example, they commonly invite rejection, abuse, and emotional distance. Parents feel rejected, unappreciated, and inade-

quate and often respond by either being punitive and rejecting, becoming depressed and withdrawn, or failing to set appropriate rules, limits, and boundaries. The situation becomes even more confusing and complicated when parents also have histories of attachment-related problems. For example, an adoptive mother, with an early history of maltreatment and abandonment, has a strong need for affection and love from her child. However, her child is angry and rejecting, thereby creating negative patterns of mutual hostility, mistrust, and emotional distance.

2. *Control battles.* Children with severe attachment disorder maintain an excessive control orientation toward parents and others. They believe that their very survival depends on coercion and manipulation. Parents, mental health professional, teachers, and extended kin often lack the information and skills required to effectively manage these coercive children. Family interaction is dominated by chronic power struggles and control battles. A substantial amount of treatment time is spent teaching effective parenting concepts and skills, with a specific focus on managing the control-oriented child.

3. *Triangulation.* Children with attachment disorder commonly form coalitions with one person against another. For example, the child will be hostile and oppositional with mother during the day and superficially charming and cooperative in the presence of father. The father sees a frustrated, angry wife and assumes she is the source of the child's problems. The wife feels criticized and misunderstood, which leads to severe marital conflict and the lack of a united parental team. Collaborative alliances must be created and maintained with parents, therapists, social services, school, and extended kin in order to avoid the negative impact of triangulation by the child.

4. *Social isolation and lack of support.* Parents routinely report feeling alone, misunderstood, blamed, and unsupported by other family members, their community, and the child welfare system in general. For example, social services and adoption agencies often place children with histories of severely compromised attachment with adoptive parents without adequate information, training, and support. These well-meaning but ill-prepared parents have "nowhere to turn" when the child begins to act-out. Another example is when grandparents, who lack understanding about the nature of the child's attachment problems, blame the parents and undermine their authority. Parents feel isolated and alone at a time when they need empathy and support the most.

5. *Marital stress and conflict.* Well-functioning marital relationships are chronically challenged by the ongoing stress of managing these children. Parents need guidance and support in order to maintain effective communication and problem-solving abilities as well as intimacy and affection. However, when there is a history of significant marital conflict prior to adoption, the added stress created by the attachment disordered child produces a

breakdown in marital functioning and commitment. Treatment must enhance the strength and unity of the marriage by addressing family-of-origin issues and improving communication patterns, conflict resolution skills, and emotional closeness.

6. *Sibling relationships.* Sibling conflicts, although naturally occurring in all families, are amplified in a family with a child with attachment disorder. These children are typically abusive, manipulative, and rejecting toward siblings. Although they reject love from all family membes, they are jealous and resentful of siblings who are capable of giving and receiving love. Siblings commonly develop secondary traumatic stress symptoms (e.g., anxiety, hypervigilance, sleep disorders, psychosomatic disorders) as a result of the victimization and ongoing stress in the family. They may feel neglected due to the inordinate amount of time and energy parents devote to the "problem" child. Feelings of guilt and shame are common because they want to love and support their siblings, but are angry and mistrustful. Their family and social life is restricted. Enjoyable family activities are curtailed, and they no longer invite friends to the home, fearing embarrassment.

TREATMENT: CORRECTIVE ATTACHMENT THERAPY

Understanding and assessing the family system when treating children with attachment disorder is crucial to treatment success. Since compromised and disrupted attachment occurs in the context of the family, not just the caregiver–child relationship, treatment must also focus on the family. Byng-Hall (1995a) describes a "secure family base" and delineates family situations that undermine family security: (1) loss of an attachment figure; (2) one family member forming an exclusive coalition with the attachment figure ("capturing"), thereby preventing a connection with the child; (3) turning to an inappropriate attachment figure due to the unavailability of an appropriate figure; (4) ongoing and destructive conflict between caregivers or caregivers and children; and (5) caregiver's unresolved attachment conflicts. Many of these problematic dynamics and patterns are present in the adoptive families we treat.

Structure and Goals of Treatment

The treatment program is called the "Two-Week Intensive" and is designed to work with children diagnosed with Reactive Attachment Disorder of Infancy or Early Childhood (DSM-IV, 313.89). Children commonly have concurrent diagnoses of Oppositional Defiant Disorder, Post-traumatic Stress Disorder, Attention-Deficit/Hyperactivity Disorder, and Depression. All suffer from a grief reaction to significant losses early in life.

Treatment occurs Monday through Friday, 3 hours per day, for 2 weeks. Families stay at local lodges. The time in therapy is divided between working with the child, the parents, and the entire family unit. A cotherapy team is utilized, providing flexibility in treatment format and focus. Follow-up therapists are encouraged to participate during treatment, which provides supervision and facilitates more effective follow-up.

Treatment always involves the child and parents. The parents are either in the therapy room with their child or observing treatment on a TV monitor in another room. The treatment program focuses on five primary areas within the family system: (1) *Child:* address prior psychosocial trauma and disrupted attachment and improve internal working model (belief system) and prosocial coping skills; (2) *Parent–child relationship:* facilitate secure attachment patterns, including trust, emotional closeness, and positive reciprocity; (3) *Family dynamics:* modify negative patterns of relating, enhance stability, support, and positive emotional climate; (4) *Parents:* address family-of-origin issues that inhibit effective personal and interpersonal functioning; and (5) *Parenting skills:* learns the concepts, attitudes, and skills of Corrective Attachment Parenting.

Initial Phase of Treatment

The beginning of treatment focuses on the parents' expectations, mind-set regarding their child and circumstances, the development of treatment goals and contracts, and establishing a working alliance and rapport. Parent often have unrealistic expectations regarding treatment outcomes and the process of change. They are informed that positive change is an ongoing process; although certain changes can occur in the short-term, more significant change solidifies over time. Parents are also often unrealistic about the focus of treatment (e.g., "fix my child"). Rather, the emphasis is on the elements of the entire family system: family dynamics and structure, parenting attitudes and skills, marital relationship, family-of-origin issues of parents, siblings and extended kin, external social systems. Parents must understand that in order to help their child they must be willing to examine and modify their own personal and relationship issues.

A primary goal is to join with the parents; i.e., convey the message that we understand the nature of their frustration. Parents are often blamed by other family members and child welfare professionals for their child's ongoing problems. The therapist must provide empathy and support to create a working alliance. This support and validation is an essential framework for the confrontation and self-exploration required for changes in self, relationships, and the family system.

Finally, it is important that parents conclude this initial phase of treatment with an enhanced sense of hope, investment in the treatment process, and realistic expectations. A specific contract between parents and therapists is established, which defines desired goals and outcomes. A recent study of families with children having severe emotional disorders found that caregivers want practitioners to give them hope (McDonald *et al.*, 1999). While it is important to caution against giving caregivers unrealistic expectations, conveying hope helps break the cycle of negativity and pessimism commonly found in the parent–child relationship.

Developing Secure Attachments

The most favorable developmental period for children to form attachments is from birth to 3 years of age. However, if secure attachment is not learned during that phase it can be learned later, although it is a more difficult task, often requiring effective therapeutic and parenting interventions.

Developing secure attachment with adopted children involves the same three processes as with biological children: first-year attachment cycle, positive interaction cycle, and claiming. In the first-year attachment cycle the caregiver gratifies the child's needs in a sensitive, appropriate, and consistent manner, resulting in a reduction of anxiety and discomfort and feelings of security, safety, and trust. Parents initiate the positive interaction cycle, creating an ongoing positive reciprocal relationship in which the child learns to respond favorably to feelings, messages, limits, and behaviors. Claiming helps the child feel "a part of" the family, a sense of belonging.

First-Year Attachment Cycle

Reciprocal interactions are infant initiated during the first year of life, constituting an ongoing attachment cycle. This cycle (see Fig. 1) begins

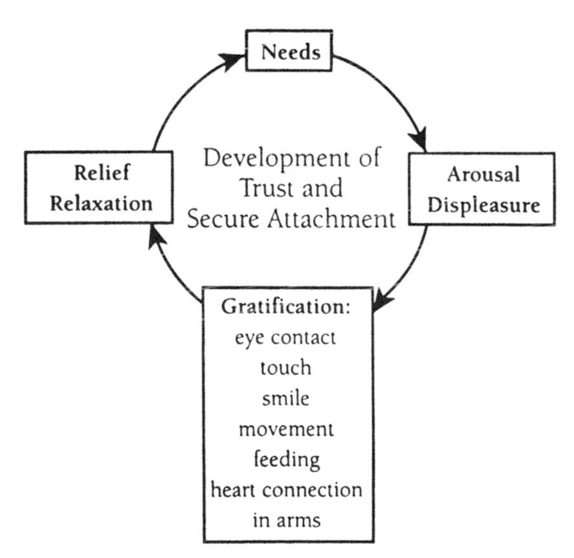

FIGURE 1. First-year attachment cycle.

with the infant's needs and arousal, is followed by the caregivers' response of gratification, and results in the development of trust and secure attachment.

Adopted children with histories of severely compromised and disrupted attachments did not experience a positive attachment cycle with reliable and sensitive caregivers. Thus, an opportunity is provided in therapy for the child to experience this cycle, first with the therapist and then with the parents. This intervention incorporates both didactic and experiential goals. It educates the parents about the etiology of attachment disorder, which enhances their understanding and empathy. It also provides the child with a cognitive frame, a way to understand his or her own thoughts, feelings, and behavior. The experiential component of this intervention utilizes the Holding Nurturing Process (HNP), an "in arms" experience between parents and child that promotes secure attachment (Levy & Orlans, 1998). The primary goal of this intervention is to enhance trust, safety, and emotional connection by facilitating a positive completion of the first-year attachment cycle. For example, the child expresses anger and frustration while maintaining eye contact and safe physical containment with an empathic parent. The child experiences a reduction of anxiety and an increase in relief and relaxation while in close physical and emotional proximity to the parent.

Positive Interaction Cycle

This intervention is parent initiated. Parents are taught to be proactive rather than reactive, to create a healing environment for the child, and to initiate a positive reciprocal relationship. Adoptive parents did not create the child's original attachment problems, but are responsible for creating an environment conducive to positive change and effective solutions. This is accomplished by the parents setting a positive family tone rather than allowing the child to set a negative family tone. Parents must take the lead by creating clear and consistent structure in order to reduce control and manipulation by the child. Children learn trust, respect, and self-control in the context of appropriate parental structure, rules, and limits. For example, attachment-disordered children commonly coerce parents into negative exchanges (escalating aggression, hopeless, and depressive climate). Parents are encouraged to maintain "calm in the face of the storm," offering consequences rather than punitive reactions. The dual emphasis on learning specific parenting skills and addressing personal emotional issues is necessary so that the parents an initiate the positive interaction cycle.

Claiming

Children with histories of maltreatment and compromised attachment typically feel isolated, alienated, and disconnected due to prior unresolved separations and losses. The child must learn to feel a sense of connection and belonging to his or her adoptive parents and family. The child must "claim" the parents and the parents must "claim" the child to create a family identity. Claiming is facilitated in several ways. First, during treatment, as the child resolves his or her grief with the adoptive parents, a heightened sense of belonging occurs. For example, as the child expresses feelings of grief regarding the loss of the birth mother, increased understanding, trust, and emotional attachment develops in the current parent–child relationship. Second, family rituals also facilitate a sense of belonging. For example, the therapeutic ritual referred to as the "inner child metaphor" is utilized. This task allows the child to tell his or her parents the "story" of what happened in the early years. Sharing memories and emotions associated with prior losses and trauma, in a context of empathy and honesty, encourages a positive parent–child connection and sense of belonging. Parents are encouraged to initiate many other ongoing rituals which help the child feel a part of the family (e.g., dinner-time, bed-time, recreation, life-books, family photos, rites of passage).

Parents who are attempting to facilitate secure attachments with children who have a history of unresolved loss and severely compromised attachment must have a variety of approaches available. Moss (1997) outlines a number of techniques.

1. Claiming behaviors such as putting the child's pictures on the refrigerator or introducing the child to family members and friends.
2. Initiating positive interactions such as an unearned hug or treat or playing a game.
3. Responding to the arousal/relaxation cycle by reacting in a nurturing way when the child is highly emotional—either due to anger or fear or sadness.
4. Encouraging physical proximity by sitting together or hugging or patting a shoulder or playing a game that requires closeness.
5. Any reenactment or early developmental tasks such as feeding, singing, or engaging in reciprocal interactions.
6. Maintaining eye contact and touching.
7. Teaching a child how to do something fun.

CORRECTIVE ATTACHMENT PARENTING

Specialized parenting skills are required in order to successfully manage and relate to angry, mistrustful, and traumatized adoptive children with attachment disorder. The primary concepts, skills, and goals of Corrective Attachment Parenting are listed below.

1. *Parents' background:* Parents' attachment histories play a significant role in their current lives. They must be aware of how prior family-of-origin issues influence their parenting attitudes and practices, marital relationships, and current psychosocial functioning.

2. *Attachment begins with the parents:* Parents are responsible for creating a framework of love, sensitivity, empathy, caring, security, and protection. They must model effective communication, coping and problem-solving skills, and management of emotions.

3. *New ideas and skills:* Parenting concepts and techniques that are effective with many children fail miserably with attachment-disordered children. Parents must be willing and able to learn totally new concepts and techniques of parenting that are effective with their own children.

4. *Parenting for attachment:* Effective parenting with attachment-disordered children must provide the same key ingredients as a secure parent–infant attachment. Parents much provide a balance of structure and nurturance which changes based on the developmental needs and capabilities of the child.

5. *The "Four R's":* Parents are taught that children are expected to be *responsible respectful, resourceful,* and *reciprocal.* Children are held accountable for their choices and actions and for responsibilities as a family member (e.g., chores, participation).

6. *Support:* Parents must have sufficient support from both inside and outside of the family. A united front is crucial in the parental team as is support from extended family.

7. *Hope:* After years of unresolved conflict and failed attempts to remedy the problem, many parents feel hopeless, demoralized, and burned out. The parenting framework must instill and enhance a sense of hopefulness, enabling parents to experience success.

8. *Specific parenting goals and skills*

- Creating a healing environment
- Providing clear and consistent structure
- Caring for self and partner
- Communicating effectively
- Providing choices and consequences
- Increasing family participation
- Parenting creatively
- Parenting that is competency based

9. *Basic objectives of effective parenting (goals for children)*

- Develop the capacity to form secure attachments and reciprocal relationships; the ability to give and receive love and affection.
- Develop the internal resources necessary to make helathy choices, solve problems, and manage adversity effectively.
- Cultivate a postive and realistic sense of self and self-in-relation to the world.
- Learn to identify, manage, and express emotions in a constructive manner.

- Learn prosocial values and morality as well as the self-discipline and self-control necessary to function successfully in society.
- Develop the capacity for joy, playfulness, and a positive meaning in life.

FAMILY INTERVENTION SPECIALIST

For many years it has been customary for some treatment programs to place children in therapeutic foster homes during intensive therapeutic intervention. There are several reasons for such a placement. First, the adoptive parents were often demoralized and burned out and in need of a break from the ongoing stress of living with an extremely oppositional and hostile child. Second, the therapeutic foster home offered a highly structured milieu with skilled and confident treatment parents. Third, as the child responded positively to the structure with more manageable behavior, parents were expected to become more hopeful and optimistic. Last, the therapeutic foster parents provided role models of effective parenting skills and attitudes for the adoptive parents.

Over time we came to realize that the above model had serious drawbacks. There was a lack of sufficient opportunities for "hands on" practice of effective parenting skills. Parents need repeated rehearsal of specific child management strategies and skills, under the guidance of a trained professional, to gain mastery and confidence. Second, respite provides only a temporary solution to a long-term problem. Parents need concepts and skills of Corrective Attachment Parenting to reduce and avoid burnout. Removing the child from their care during treatment merely provides a reprieve from the stress, not a long-term solution. Third, when the child did better in the treatment home, the covert message to the child was: "others can manage me but my parents cannot." Lacking direct experience with their own parents being effective, children did not develop confidence in their parent's abilities to manage them. Last, many parents reported feeding inadequate in comparison to the trained therapeutic foster parents. Thus, rather than feeling increased confidence and optimism, they felt ill-prepared for the task at hand. The covert message the parents received was: "others can manage our child be we cannot."

A significant correlation was found between treatment failures and the lack of adequate Corrective Attachment Parenting skills. Despite the children's improvements while in the treatment home, serious difficulties occurred upon reunification with their adoptive families. In an attempt to refine the treatment process—maintaining the benefits while reducing the drawbacks—we developed the concept of the *Family Intervention Specialist.* The primary role and goals of the Family Intervention Specialist are listed below.

1. *Education:* A primary task is to provide information and facilitate skill-building. The Family Intervention Specialist imparts information regarding the specifics of attachment disorder and Corrective Attachment Parenting. For example, an explanation is given to the parents regarding their child's oppositional and controlling behavior. The parents can then understand the importance of being proactive rather than reactive in order to avoid power struggles and control battles. Parenting skills are discussed and practiced under the guidance of the Family Intervention Specialist (e.g., consequencing; managing emotion, stress, and resistance; communication and problem-solving skills).

2. *Support:* As previously discussed, parents who come for treatment are typically demoralized and highly stressed. Another crucial responsibility of the Family Intervention Specialist is to supply much needed emotional and moral support to all family members, especially the parents. Support helps the parents feel understood, builds their confidence, and is empowering. For example, parents who are commonly blamed and disempowered by extended kin and well-meaning child welfare professionals (who are unaware and/or misguided regarding attachment disorder) feel relieved and empowered by the advocacy role of the Family Intervention Specialist.

3. *Tasks:* During the first day of treatment the Family Intervention Specialist intervenes directly with the child and also serves as a member of the treatment team with the parents. For example, the child is given a sentence completion form and pretreatment evaluation assignment. The Family Intervention Specialist observes the child's reactions to the tasks and deals with questions, resistance, and other patterns of reacting. Throughout the treatment process the Family Intervention Specialist assigns specific family tasks (e.g., chores, rituals), and serves as a "coach" in the successful completion of these tasks. The Family Intervention Specialist does not engage in intensive therapeutic interventions; it is important to maintain clear boundaries.

4. *Crisis management:* Crisis intervention can occur in the clinic, at the families residence, or via telephone. The primary goal is to facilitate positive changes in family dynamics and patterns of coping and reacting. The Family Intervention Specialist provides information and support and encourages effective problem-solving skills in order to deescalate the crisis. For example, the Family Intervention Specialist provides the parents with effective strategies to handle their child's noncompliant and aggressive behavior.

5. *Treatment team:* The Family Intervention Specialist is a member of the treatment team and serves as a liaison between the therapists and the family. The Family Intervention Specialist can provide the therapist with valuable information regarding individual and family dynamics observed beyond the treatment setting. Additionally, the Family Intervention Specialist reinforces and encourages issues and goals addressed in therapy. For example, a triangulation process is identified during treatment in which mother and child are in a power struggle and father is unsupportive of his

wife. Later that day, the Family Intervention Specialist encourages the father to support to his wife during a family conflict, thereby providing a united parental team to their child.

6. *Follow-up:* The primary therapist and the Family Intervention Specialist have specific and different follow-up responsibilities. The Family Intervention Specialist provides follow-up consultation regarding practical, day-to-day parenting issues. The therapist is responsible for all other treatment follow-up concerns (e.g., consults with hometown therapist, school, and social service caseworkers).

Although the therapeutic foster home is not typically used as a primary treatment modality, there are circumstances that warrant the placement of a child in a therapeutic home during treatment. For example, a child who is particularly oppositional, controlling, and resistant to treatment can be placed in a therapeutic foster for a brief period of time (e.g., 1 or 2 days). This is a paradoxical intervention; rather than trying to convince the child to participate in therapy, the child must now demonstrate that he or she wants to earn therapy back. The vast majority of children do demonstrate increased motivation and genuine involvement in the treatment process once they are not successful at maintaining their control orientation.

CONCLUSIONS

Children with compromised and disrupted attachment enter the adoptive family with an array of prior psychosocial symptoms and negative relationship patterns. The family, in many cases, is literally importing pathology into the system. It is not sufficient, however, to focus only on the child. It is necessary to understand the child within the context in which he or she functions—the family. The family systems approach focuses on assessing and changing relationship patterns. Family relationships are reciprocal and circular, with the behavior of each family member serving as both a trigger and response for one another. These ongoing patterns of interactions maintain the family system and often the behaviors and symptoms of the adopted child with attachment disorder. For example, when a child with a history of maltreatment, several out-of-home placements, and anxious and/or disorganized attachment is adopted into a family, the focus becomes not only the child's history of problems, but also the constellation of family-related issues: parents' attachment histories, marital relationship issues, sibling issues, parenting attitudes and skills, relationship patterns and dynamics, and external social systems.

Many parents who adopt special-needs children, such as those with attachment disorder, are psychologically capable of meeting the challenge. Others, however, have histories of dysfunctional family relationships as well as current individual and marital difficulties. In these cases, treatment must focus, in part, on the parent(s) issues to avoid scapegoating the child. The therapist must walk a tightrope to maintain a delicate balance. On the one hand, provide empathy and support to parents who are feeling disgruntled, hopeless, and blamed. On the other hand, confront the parents' own issues to effect necessary change.

REFERENCES

Berman, P., Grekin, E., & Mednick, S. (1999). Maternal smoking during pregnancy and adult male criminal outcomes. *Archives of General Psychiatry, 56*(3), 215–224.

Besharov, D. J. (1994). *When drug addicts have children.* Washington, D.C.: Child Welfare League of America.

Borysenko, J., & Borysenko, M. (1994). *The power of the mind to heal.* Carson, CA: The Hay House, Inc.

Bowlby, J. (1969). *Attachment and loss: Attachment* (vol. 1). New York: Basic Books.

Bowlby, J. (1970). *Child care and the growth of love.* Harmondsworth: Pelican.

Brazelton, T. B., & Cramer B. G. (1990). *The earliest relationship.* New York: Addison-Wesley.

Brodzinsky, D. M., & Schechter, M. D. (1990). *The psychology of adoption.* New York: Oxford Univ. Press.

Brodzinsky, D., Schechter, M., & Henig, R. (1992). *Being adopted: The lifelong search for self.* New York: Doubleday.

Byng-Hall, J. (1995a). Creating a secure family base: Some implications of attachment theory for family therapy. *Family Process, 34*(1), 45–58.

Figley, C. (1989). *Helping traumatized families.* San Francisco: Jossey-Bass.

Figley, C. (1998). *Burnout in families: The systemic cost of caring.* New York: CRC Press.

Jones, A. (1997). Issues relevant to therapy with adoptees. *Psychotherapy, 34*(1), 64–68.

Levy, T., & Orlans, M. (1998). *Attachment, trauma, and healing: Understanding and treating attachment disorder in children and families.* Washington, D. C.: Child Welfare League of America.

Lindholm, B. W., & Touliatos, J. (1980). Psychological adjustment of adopted and nonadopted children. *Psychological Reports, 46,* 307–310.

McDonald, T. P., Poertner, J., & Pierpont, J. (1999). Predicting caregiver stress: An ecological perspective. *American Journal of Orthopsychiatry, 69*(1), 100–109.

Moss, K. (1997). *Integrating attachment theory into special needs adoption.* Cleveland, OH: Beech Brook.

Samuel, M., & Samuel, N. (1986). *The well pregnancy book.* New York: Simon and Schuster.

Sontag, L. W. (1970). Parental determinants of postnatal behavior. In H. Weisman & G. Kerr (Eds.), *Fetal growth and development* (pp. 265–281). New York: McGraw–Hill.

Verny T., & Kelly J. (1981). *The secret life of the unborn child.* New York: Delta.

Watson, K. (1997). Bonding and attachment in adoption. In H. Gross & M. B. Sussman (Eds.), *Families and adoption* (pp. 159–173). New York: Haworth Press.

BIOGRAPHIES

Terry M. Levy

Dr. Levy is a Licensed Clinical Psychologist in Colorado and Florida and a Board Certified Forensic Examiner in Clinical and Family Psychology. He is a clinical member of the American, Colorado, and Florida Psychological Associations, American and Colorado Association of Marriage and Family Therapy, American Family Therapy Academy, American College of Forensic Examiners, American Psychotherapy Association, and the National Register of Health Service Providers in Psychology. He was founder and Director of the Family Life Center (Florida) and the Miami Psychotherapy Institute, and cofounder and President of the Board of Directors of the Association for Treatment and Training in the Attachment of Children (ATTACh). Dr. Levy has been providing psychotherapy treatment and training for over 25 years. He has taught seminars on therapeutic issues for the American Psychological Association, the American Professional Society on the Abuse of Children, and numerous mental health, social service, and school systems nationwide. Currently, Dr. Levy has a private practice with Evergreen Consultants in Human Behavior. Dr. Levy is coauthor of *Attachment, Trauma and Healing* (1998, Child Welfare League of America).

Michael Orlans

Michael Orlans is in private practice and Director of Training with Evergreen Consultants in Human Behavior. He is the former clinical supervisor for the Domestic Intervention Program—State Attorney's Office, 11th judicial circuit of Florida. He is a certified criminal justice specialist and is a Marriage and Family Therapist with over 25 years of clinical experience. In addition to his work in private practice and in the public sector, Michael has taught on the faculty of several universities. He is a nationally known lecturer and trainer, with expertise in working with severely emotionally disturbed children and their families. He has served as a consultant to therapeutic foster care programs, child welfare agencies, and is on the Advisory Council of the National Alliance for Rational Children's Policy. He is cofounder of the Association for Treatment and Training in the Attachment of Children. He is also a Diplomate, Board Certified Forensic Examiner, and founding executive board member and Diplomate of the American Psychotherapy Association. He is also a coauthor of *Attachment, Trauma and Healing,* Child Welfare League of America Press.

10

COMMUNITY-FOCUSED ATTACHMENT SERVICES

PAULA PICKLE

The Attachment Center at Evergreen, Inc.
Evergreen, Colorado 80437-2764

INTRODUCTION

Attachment disorder, unlike many other mental or behavioral disorders, occurs in the context of relationships. Attachment disorder, the inability to form loving, lasting intimate relationships, is demonstrated through a lack of reciprocal behavior, violations of the rights of others, frequently aggressive and destructive acts, and a lack of remorse for behavior that is hurtful to others. Those factors which place a child at risk for developing an attachment disorder primarily occur in the context of the parent–child relationship. The ability of a parent to provide essential care to an infant is also based on relationships. Family relationships can be enhanced or diminished by the fabric of our society.

With an increasing awareness of the risk factors for the development of attachment disorder and an increasing awareness of the difficulty in successfully treating a child with attachment disorder, it makes sense for society to begin to address those risk factors. Prevention of attachment disorder should be our primary goal. Successfully treating a child with attachment disorder requires a coordinated team approach, enlisting the assistance of significant others and community systems. Attachment disor-

Handbook of Attachment Interventions

der is not addressed by focusing only on the child. Failure to recognize the severity of this problem or to address it from a societal level, has serious consequences at all levels of our society. The costs of ignoring this problem are great, both on a financial level and in terms of human lives. This chapter will explore these issues and will describe a program which incorporates some of these principles. This chapter will also highlight some steps which could be taken to reduce the impact of attachment disorder on individuals and on society.

ETIOLOGY OF ATTACHMENT DISORDER

Many of the original studies exploring the effects of a child's separation from parents found similar reactions in children. This study of the relationship between child and parent led to an understanding of the importance of attachment for normal child development. "A young child's experience of an encouraging, supportive and co-operative mother, and a little later father, gives him a sense of worth, a belief in the helpfulness of others, and a favorable model on which to build future relationships. Furthermore, by enabling him to explore his environment with confidence and to deal with it effectively, such experience also promotes his sense of competence" (Bowlby, 1982).

The factors that lead to a breakdown in the attachment between parent and child can be grouped into three major areas: biological factors, familial factors, and societal factors.

Biological Factors

In Utero Exposure to Alcohol or Drugs

In utero exposure to alcohol and drugs has long been known to damage the developing fetus in numerous ways. Children whose neurological functioning is damaged prior to birth have a difficult time with ordinary processing tasks. This makes it difficult for them to accomplish the normal developmental tasks and to make the connections which are part of the fabric of human experience. A child with sensory integration problems or tactile defensiveness is not responsive in a positive manner to the nurturing touch and cradling which is necessary for attachment to occur.

Maternal Ambivalence Toward the Pregnancy

Maternal ambivalence toward the pregnancy is another *in utero* factor which significantly affects the developing child. Studies cited by Thomas Verny emphasize the importance of maternal attitudes toward the pregnancy for healthy child development (Verny, 1981).

Difficult Birth

A difficult or traumatic birth can adversely affect the relationship between mother and child. The ordinary process of labor and the struggle to be born culminates in the relief and bonding that occur as a mother and her child come face-to-face for the first time. Births which are crisis oriented—with potential threats to the health or life of mother or child—frequently result in mother or child being removed in order to provide medical care. This can result in a barrier to the normal bonding which occurs through the birth process.

Undiagnosed and Untreated Illness or Severe Medical Problems

Severe medical problems with the child or mother frequently results in a separation of the child from the mother. This separation is one of the risk factors for development of attachment disorder. Pain which the parent cannot relieve interrupts the normal trust cycle (need → cry-for-help → relief → trust) which forms the basis of attachment.

Child's Innate Temperament and Intelligence

A child's natural temperament and intelligence also play a role. A study conducted by Tim Harlan recently found that children with certain personality traits were more likely to develop attachment disorder than those with other traits and that children with higher intellectual functioning were more likely to develop attachment disorder (Raudolph, 1997).

The Developing Brain

New research has shown the extreme importance of the first years of life, including the prenatal period, for the development of the brain. These early life experiences have a precise impact on how the intricate neural circuits of the brain are wired. Failure to provide a consistent, nurturing, stimulating environment has dire consequences for a child's brain development.

Familial Factors

Generational Factors or Legacies

Generational factors or family legacies have a significant impact on the quality of the parent–child relationship. Most individuals are likely to parent in ways similar to the way they were parented. Abuse begets abuse. Emotionally withdrawn parents shape children who are uncomfortable with intimacy. Children growing up in chaotic families have chaos as their point of reference for parenting.

Ill-Prepared Parents with Poor Parenting Skills

Parents who are unprepared to parent or who have poor parenting skills also place a child at risk for developing attachment disorder. Children require consistent and nurturing attention to their needs in order to develop a secure attachment. Their sense of self, security, ability to regulate emotions, and ability to trust and to develop cause-and-effect thinking depend upon their caregiver's ability to meet their needs. The numbers of children bearing children is staggering. These immature parents cannot meet their own needs let alone the needs of an infant. In 1995, 512,115 babies were born to teenage girls; 12,242 of these were to mothers younger than 15 (Children's Defense Fund, 1998).

Maternal Depression

Maternal depression plays a significant role in the emotional health and well-being of an infant. While a depressed mother may meet the physiological needs of an infant, the emotional connection and the mother's attachment behavior may not be enough to communicate to the child how much he/she is loved or cared for. Children of depressed mothers are at much greater risk for emotional difficulties during childhood and adolescence (Zuckerman & Beardslee, 1987).

Childhood Abuse, Neglect, and/or Abandonment

Childhood abuse, neglect, and/or abandonment are significant factors in the development of attachment disorder. These are so disruptive to the normal bonding/attachment cycle because they communicate to the child the message that he/she is not loved, not worthy of care, and powerless to change. While many children with backgrounds of abuse/neglect/abandonment do not grow up to have attachment disorder, this background has severe consequences for their sense of self-worth and their ability to trust. Many children with these backgrounds do grow up to continue the cycle of abuse. In 1995, there were 2.96 million reports of child abuse/neglect, 1.11 million of these were substantiated. Of these there were 565,000 serious injuries and 996 fatalities. This represents an increase of 42% in 10 years (Petit, 1997).

Societal Factors

Alcohol and Drug Abuse

Alcohol and drug abuse contributes to many of the factors listed above. Chaotic families, abuse and neglect, divorce, depression, children suffering the effects of *in utero* exposure, low-birth-weight babies, birth trauma, and premature births are some of the results of alcohol and drug abuse. Children born to alcohol- or drug-dependent mothers have several risk factors predisposing them to attachment disorder. Not only has there been physiological or neurological damage, but their mother is not emotionally available. There is an increased risk of abuse, neglect or abandonment. The parent may be incarcerated, necessitating the child's placement in foster care. Children of alcoholic parents are more likely to abuse alcohol and drugs themselves.

Inadequate Day Care

Day care for infants and toddlers may be necessary for a mother to be able to support herself and her child. Day care with inadequate staffing patterns (more than three infants or toddlers per staff member), high staff turnover, or staffing patterns that do not allow a child to have a consistent and nurturing caregiver increase the uncertainty for a young child and do not allow that child to form trusting relationships with adults. "Every day, 13 million children—including 6 million infants and toddlers are in child care. Two-thirds of mothers of young children work outside the home—many out of economic necessity" (Children's Defense Fund, 1998).

Teen Pregnancy

Teen pregnancy is a societal problem with severe implications for the well-being of the children born to these teens. Not only are these parents ill prepared for the task of parenting, many have emotional problems of their own. It is unlikely that these parents will be able to complete their education, find meaningful employment in order to support themselves and their child, and find adequate child care.

Single-Parent Heads of Households

Single parent heads of households are a growing segment of our society. More than a quarter of all children live with one parent—usually the mother. Almost half of all children today can expect to experience a divorce during childhood and to live an average of 5 years in a single-parent family (Carnegie Corporation, 1994). A single parent, as the sole wage earner for the family, has all the household obligations and financial burdens for the family. In addition, they frequently have little emotional energy left to take care of children or himself/herself. Single parents many times are economically disadvantaged as well, which can make adequate child care unaffordable.

Breakdown of the Family

The breakdown of the family is also a significant contributing factor. Regardless of the specific cause for the family disruption, the end result can include parental depression, economic difficulties, nonavailability of one or both parents, a sense of abandonment on the part of the child, chaotic family situations, moves, and child care changes, to mention a few.

These results are incompatible with a child's needs for consistent, available, and nurturing care.

An Overburdened Child Welfare System

An overburdened child welfare system, whose task is to protect the most vulnerable children, frequently is unable to meet the needs of the children it serves. Children removed from their families because of neglect, abuse, or abandonment need the best that can be provided. Because of their early life experiences, they may already have difficulty trusting caregivers. All too frequently, these children are difficult to parent and end up being moved from foster home to foster home while the system slowly works out a permanent plan for the child. Thus the system, whose purpose is to protect these children, frequently compounds the problem. Enormous caseloads, staff that are untrained to recognize emotional problems, delays in getting needed help, and lack of timely permanency planning for the children are contributing to the growing numbers of children with attachment problems.

Adoption and Foster Care

Adoption and foster care are necessary for the care of children whose life circumstances make it impossible for them to live with their birth families. However, a thorough understanding of the effects of this on children and on their adoptive or foster parents is essential if we are to place children in a planful way. Our system has too often acted in ways that were not in children's best interests. Some problem areas are listed below.

1. Prolonged delays in permanency plans.
2. Unspecific requirements for parents to retain parental rights.
3. Lack of matching child and families.
4. Failure to disclose important information to adoptive or foster parents.
5. Faulty screening of adoptive or foster parents.
6. Failure to adequately evaluate the child's needs for specific services.
7. Failure to provide specialized services for children.
8. Failure to demand that mental health service providers have experience and training in the areas of adoption/foster care/abuse/neglect/attachment issues.
9. Failure to adequately support foster and adoptive parents experiencing difficulty with the children placed with them.

Our system has not adequately prepared itself to choose the best interests of the child in custody disputes, challenged adoptions, or other critical decisions relative to child placements. One need only recall seeing on national television young children tearfully removed from adoptive families and returned to birth families. Legalities and parental rights take precedence over the child's primary attachments.

COMMUNITY ISSUES

Complicating Factors

Unrecognized Disorder

Attachment disorder has been relatively unrecognized as a severe and legitimate disorder. The Diagnostic and Statistical Manual of Mental Disorders utilized by mental health professionals has a description of this disorder which is inadequate. Reactive Attachment Disorder of Childhood (DSM-IV) does not reflect what those professionals who treat children with this disorder have come to recognize as a legitimate and predictable cluster of symptoms. Therefore many children are improperly diagnosed. If it is true that a diagnosis suggests appropriate treatment, then failing to properly diagnose a child can lead to inadequate treatment.

The importance of attachment has been known since at least the 1950s, yet this is not considered an essential part of college course work in schools that are preparing the professionals who are called upon to provide services to children with this disorder. School systems, medical professionals, social service agencies, and other important community systems who are uninformed about this disorder can inadvertently undermine the parents or therapists who are working with the child who has this disorder.

Limited Number of Trained Resources

There is a limited number of professionals who have a good understanding of attachment disorder and who have the training and expertise to effectively work with the child who has attachment disorder and his/her family. Conventional therapy is ineffective in addressing these issues for several reasons. Therapeutic relationships are based upon mutual trust and respect and the ability to be emotionally honest. Therapy, which is largely a cognitive process, will not touch the deeply emotional roots of attachment disorder. Those professionals who work with character-disordered adults know the futility of effecting change in individuals who have no internal conflict concerning their ways of relating to others. Many parents have spent years and thousands of dollars seeking help from professionals who have been unable to bring about significant change in the child's behavior or the parent's understanding of the problem.

Lack of Understanding

School systems, social services, and medical providers who do not understand or recognize attachment disorder can inadvertently undermine the efforts of parents or therapists who are working with a child with attachment disorder. It is not uncommon for these systems to feel that they must "protect the child" from the parents. Children with attachment disorder are very adept at triangulating adults or manipulating others into blaming parents. Rescuing a child under these circumstances might result in a change of placement for the child. This further contributes to the child's attachment problems and does irreparable damage to the parent–child relationship.

Unprepared Foster and Adoptive Parents

While it is extremely difficult to prepare adoptive or foster parents for the experience of parenting a child with attachment disorder, it is essential that these parents are provided with all the background information on the child prior to placement. Parents must be able to make informed consent as to their ability to parent a child with the types of behaviors which the child being placed exhibits. Failure to inform, for the purposes of securing a placement for a child, is unethical behavior because it sets up a child and a family to fail. In addition to providing essential information for the adoptive or foster parent, it is also important to provide the necessary financial support and follow-up services to help these parents succeed with the child. All too often, children with severe problems, such as attachment disorder, are placed with parents without adequate information or preparation. In order to access specialized services, these parents must frequently take on the entire system and many times these parents are blamed, blocked, or delayed in their attempts to care for their child. The emotional and financial costs of these battles are overwhelming.

Curative Factors

There are several curative factors that mean the difference between success or failure when working with a child and family struggling with attachment disorder.

Tight Teamwork

A tight team is one of the key elements for success. One-on-one therapy with the child, which excludes the parents, is counterproductive. It allows the child to be less than emotionally honest and to triangulate between the adults. Therapist and parents must join as a team to develop a plan that safely contains the child's behavior and allows the child to learn better ways to relate in his/her family. Other team members who must be on the same page include physicians, school teachers, social service workers, extended family, friends, neighbors, or anyone else who may have occasion to interact on a regular basis with this child or family. Because a child with attachment disorder avoids dealing with his or her important emotional issues by controlling the environment and keeping others, particularly parents, at a distance, it is essential that important adults in his/her life consistently respond in ways that allows the child to learn from his/her errors in judgment. Anytime the team is divided in it's responses, the child suffers and progress is lost.

Continuum of Supportive Services

Effectively treating a child with attachment disorder requires a continuum of services. Treatment by trained and skilled attachment therapists is essential, but this cannot succeed without committed, healthy, trained parents who will consistently provide the necessary type of parenting. Parents cannot effectively maintain their role without trained respite care to allow time to take care of themselves, their marriage, and any other children in the family. Back-up supportive services such as residential care, psychiatric support, and financial assistance for the frequently high cost of treating these children are also important. Many children need specialized services to help with other identified needs such as sensory integration difficulties or learning disabilities.

A supportive community network that is working together for the good of the child is essential for success. Support groups can also provide emotional support as well as assistance in planning parenting and community strategies.

Prepared and Functioning Family Unit

Parents who are well informed and who are educated about attachment disorder as well as the skills to effectively parent a child with attachment disorder are more likely to be successful. "Normal" parenting will not be enough. Prepared parents are much more likely to "be there for the long haul"—an essential curative factor in itself.

In addition to education and preparation, parents should be relatively healthy mentally and emotionally. If they have unresolved issues from their own childhoods, these must become part of the equation for treatment. Children with attachment disorder are adept at knowing a parent's issues and using those issues to "push the parent away." If a parent's own issues are severe, a child does not feel safe to allow himself/herself to be vulnerable or dependent upon that parent.

Marriages have suffered severely as a result of living with a child with attachment disorder. Parents must be able to nurture and support each other in their task as parents. This is extremely difficult because the child with attachment disorder relates to each parent in a different fashion. Usually the mother is perceived as a "bad parent" while the father is related to as a "rescuer from Mom." If the marriage is not equipped to resolve these difficulties, then the child can further drive a wedge between parents. Again, the child loses the support of an effective team.

Parents must be able to commit to parenting as long as it takes. In some instances, this means parenting from a distance. It is important for a child with attachment disorder to know that these parents are committed to him/her.

Time, Energy, and Financial Resources

To successfully treat a child with attachment disorder requires a great deal of time and energy on the part of parents and therapist. This is not a job for the faint of heart. The demands frequently require more than the standard 50-minute hour for the therapist. The therapist must be ready to commit to unscheduled sessions, phone calls from parents regarding parenting strategies, and supportive sessions with the parents. The therapist will need to commit to facilitating cooperative relationships with community systems, such as the school system and social services. The therapist must be ready to commit to being a long-term therapist—frequent changes in therapists is counterproductive. Once an effective team of therapist and parents is developed, it is wise not to disrupt it. Funding for mental health services is woefully inadequate for such a severe and difficult problem. The mental health services required to address this problem do not fall neatly into the standard description of funded services and therefore accessibility to needed services is limited.

Skilled Attachment Therapists

A highly skilled attachment therapist, with a good professional support network is another key to success. Effective treatment is highly specialized, requiring training and supervision. Because of the difficult nature of the work, a group practice is recommended to provide support, back-up, and clinical review of techniques. The Association for Treatment and Training in the Attachment of Children (ATTACh) organization was established as an international group of professionals and parents dedicated to addressing the issue of attachment disorder. Annual conferences provide opportunities for additional training and networking.

Societal Strategies

Prevention

Since the risk factors for development of attachment disorder are known, these should be a focus of prevention activities. There are many ways that prevention can be effective.

1. Community education programs which emphasize the importance of attachment for normal child development can also provide training in ways to promote attachment.

2. Hospital or community programs can help to teach new mothers how to nurture and to care for their child.

3. Critical staff training can alert those who come in contact with new mothers to recognize signs of difficulty and to offer assistance.

4. Day care regulations can address the need for consistent caregiving of very young children and can stress staffing patterns and training needs of child care providers.

5. Training requirements for those entering the child welfare or mental health field must include an emphasis on attachment and it's critical importance to child development. Staff must learn to recognize attachment difficulties and know where to go for help with these issues.

6. Prenatal care emphasizing the importance of caring for self and the developing child can help promote a positive outcome for the child and the mother.

7. Community programs which focus on prevention of alcohol and drug abuse and prevention of teen pregnancies are also important.

8. Community programs that strengthen marriages and families are essential prevention programs.

Early Assessment and Intervention

Early assessment and intervention are ways that the problem can be addressed before it grows and the child's behaviors become well entrenched. Every child entering the foster care system and being placed for adoption should be evaluated for the possibility of attachment difficulties. Child welfare staff should be well trained in the area of attachment, separation and loss, and matching children to families. Therapists who work with the foster care population should have extensive and specialized training to prepare them to evaluate and to intervene with foster children and their families. All therapists must become familiar with this very specific population and be able to recognize the problem. Those who lack the specialized training necessary to work with this population should assist the family in finding the appropriate trained resources.

Social Policies

Social policies in many different areas have a tremendous effect on this important issue of parent–child attachment. Policies in the areas of day care, welfare, adoption, foster care, health care, and mental health care have far-reaching consequences for the health and well-being of our nation's children and of our society. Family leave policies and other policies affecting the strength of the family as a primary social unit must be carefully explored. Attachment should be a critical element of consideration in all policies in these areas.

IMPACT ON THE COMMUNITY

Criminal Justice and Correctional Systems

The criminal justice and correctional systems are filled with adults who have not learned to live within the boundaries established by laws and who have not learned to respect the rights of others. Many of these have backgrounds filled with the risk factors that produce children with attachment disorder. Many of them had demonstrated behavior as children and adolescents that would have predicted this outcome as adults. A high percentage of these adults have come from backgrounds of abuse, neglect, abandonment, and foster care. Their lives as adults are not productive. The lives of their victims are forever changed, some irreparably. The financial cost to society to incarcerate these individuals is staggering. When the costs of lost productivity and human lives are considered, then the costs and comparative value of prevention, early assessment, intervention, and treatment is easily seen.

Mental Health Systems

Many of our mental health systems are filled with adults who could have been diagnosed as attachment disordered as children. They are diagnosed now as character disordered. Their lifestyles are fairly well entrenched and unlikely to change unless they experience enough internal conflict to do the necessary corrective work. These individuals are usually not productive in society. Their behavior causes a great deal of discomfort and conflict for those closest to them. These are individuals who might have lived productive lives had they received appropriate treatment earlier.

Too many children with attachment disorder are treated in long-term residential care or group care or their behavior is managed through the use of pharmaceuticals. Many of these children could learn to live safely in a loving family with appropriate treatment. This is especially true if treatment is offered early in the child's life.

Homeless Population

Children with attachment disorder grow up taking what they need, feeling entitled to what they take. They have not learned to give back, to contribute to society. They would not choose to work. They do not recognize authority figures and do not like to be told to do anything. They are unlikely to be productive citizens. When they run out of others to "take care of them" they are likely to become homeless.

Alcohol and Drug Systems

Individuals with attachment disorder have spent their lives trying to avoid any emotional pain. Frequently coming from family backgrounds of alcohol and drug abuse, they are likely to turn to these substances as a means of avoiding emotions, relationships, and responsibilities. All the problems associated with the abuse of alcohol and drugs is compounded by the relational style of an individual with attachment disorder. Damage done to the developing child *in utero* when a parent uses alcohol and drugs is severe and has lifelong consequences for the child. Yet, many mothers have child after child affected with fetal alcohol syndrome or fetal alcohol effects. Society must develop a strategy that protects these vulnerable children while protecting the individual rights of the mother. Society ends up paying for the treatment or incarceration costs of the parent and the care of the children as well as assistance offered to any victims.

Public Health Systems

The public health system is one agency that interacts on a regular basis with many of those most vulnerable to develop attachment problems. Family planning offers information to teens regarding birth control and sexually transmitted diseases and well-child clinics and nutrition programs provide services to vulnerable children and economically disadvantaged parents. Public Health Systems provide health care to some of the most vulnerable children and their parents. The financial costs to society are great but this also represents a unique opportunity to provide parent education, to assess children, and to encourage early intervention if problems are noted.

Domestic Violence Systems

Domestic violence programs are filled with individuals who have backgrounds of abuse, neglect, and abandonment. These are individuals who do not have the capacity for reciprocal relationships. Their relationships are characterized by power and control as well as entitlement issues. Likewise, victims of domestic violence are frequently individuals who grew up in that role. These are individuals who have not learned to respect themselves or others. Children who grow up in these situations are more likely to have emotional and behavioral problems as well.

Welfare Systems

Welfare systems are overburdened with the responsibility of caring for parents and children who have been forced to rely on the government to provide for their needs. Adults with attachment disorder produce children. These parents have no capacity for caring for another human being, especially a child. Respect for others and responsibility for one's actions are foreign concepts to an adult with attachment disorder. Yet, these adults can be superficially charming and engaging, convincing others readily of their sincerity. They prey upon the vulnerable, leaving them to pick up the pieces of broken relationships and to care for the children they have produced.

Child Protection Systems

Child protection systems have the difficult task of protecting the most vulnerable children from abuse and preserving the sanctity of the family wherever possible. Numbers vary, but estimates of 400,000 to 600,000 children per year have been removed from their homes and are cared for in our nation's foster care system. Approximately 100,000 of these children might be available for adoption as parental rights have been terminated. Many children in foster care have several moves within that system. A child with attachment disorder is frequently moved due to the difficulty of their behavior. Ten moves is not uncommon; a number of children have upward of 20 moves. How does this frequent change in caregiver help a child develop trust, attachment, problem-solving skills, and a sense of self-esteem? At what point does the system begin to think that this child's needs are not being met by the current system or by the services being provided?

Another dilemma faced by the child protection system is the difficulty in sorting out true allegations of abuse from false allegations of abuse. Children with attachment disorder very quickly figure out how to work this system to avoid doing anything they don't want to do. Power and control, revenge and payback, manipulating and triangulating adults, and meeting their entitlement needs can all be accomplished by a well-thought-out and acted-out lie.

Pathological lying comes naturally to children with attachment disorder. Since many of these children come from backgrounds of abuse or neglect, they have the knowledge to be convincing. Sadly for these children, they think they are winning. What they are doing is adding one more loss to their history, adding more adults to their list of people they have successfully manipulated and successfully destroying adults who have dared to care for them. On the other hand, these children can also be easy prey, as they are known to be less than truthful, it is sometimes difficult for them to convince others when they are telling the truth. Discerning the truth when allegations are made requires the wisdom of Solomon.

General Well-Being of Society

A society with growing numbers of children and adults with attachment disorder becomes a weak society. The fabric of society critically depends upon relationships that are built on mutual respect, individual responsibility, and community compassion. A society that ignores the critical elements necessary for child development is a society that ignores it's own future. What affects one, affects all. The critical socializing unit of society is the family. And yet, the family is disintegrating rapidly under the onslaught of difficulties in today's world. Neglecting to address those issues affecting the nation's children and families puts everyone in jeopardy.

A COMMUNITY-FOCUSED APPROACH TO INTERVENTION

The Attachment Center Model for Treatment

The Attachment Center at Evergreen, Inc., is a treatment and training program focusing on attachment disorder. A nonprofit organization, The Attachment Center at Evergreen has been in existence since 1972. As the understanding of attachment disorder grew, so the program has evolved to address many of the issues described above.

The treatment approach at The Attachment Center begins with a thorough assessment of the child, family, and community system and then carefully assembles a treatment team that matches the child and family to a therapeutic foster parent and therapist. Treatment occurs within the context of a therapeutic family setting because this is the setting where the child with attachment disorder has the greatest difficulty functioning. Therapeutic parents team with placing parents to develop a parenting strategy that works for the particular child.

Each day of treatment, the entire treatment team meets to discuss issues, progress, and problems and to develop strategies. The focus of treatment includes a focus on family systems and individual work with parents as necessary. Therapy with the child includes contracting for treatment, addressing early experiences, educating as to the basis for his/her behavior, resolving trauma, learning how to appropriately express emotions, learning how to be reciprocal and respectful in relationships, learning how to be accountable for behavior, and correcting any thought distortions that interfere with healthy functioning. A psychiatric evaluation helps to clarify the diagnostic picture and to discern whether there is a biochemical basis for the child's behavior. A neurodevelopmental assessment helps to pinpoint problems in that area so that a program can be developed to maximize the child's ability to process information and to function.

An essential part of the treatment process is the inclusion of a hometown therapist. Since the majority of referrals to this program come from other states, it is imperative that the placing family bring a hometown therapist for a 2-week intensive treatment process. This ensures that the hometown team is well prepared to carry out the treatment plan upon their return home. Both the hometown therapist and the parents receive extensive training during the course of treatment. They develop a thorough understanding of the issues and are prepared to return home to establish a supportive community network to sustain the gains they have made through treatment.

Reintegration and follow-up services include a visit to the placing parents by the therapeutic foster parent to sustain the gains made by the child. The therapeutic foster parent frequently meets with school personnel, support groups, and respite care providers during the reintegration visit. Occasionally, they also provide community workshops or awareness seminars during this visit as well.

Replicating This Model on the Community Level

This is a model that addresses many of the issues relative to attachment disorder. Components of this model can be adapted to most community settings. This can be accomplished by a group of interested parents and professionals working together and involving the other community systems who are critical to address this particular problem. The first step is to increase awareness of attachment disorder, including its etiology and impact on society. The next step is to brainstorm about the services currently in place that might be altered or adapted to address this problem. Many groups or organizations working together can accomplish much. A coalition of individuals and agencies can develop a strategy for improving the services to this population. It is a matter of making this a priority. Communities can work together to prevent attachment disorder and to treat children with attachment disorder or they will end up working together to pay for the impact of attachment disorder on their community.

SUMMARY

In the April/May, 1998 issue of *Zero to Three,* Irving B. Harris records this story which he had originally heard from Reg Lourie, the first Chairman of the Board of Zero to Three. The story is really an allegory which accurately depicts the problems facing communities with regard to children. As the story goes, a man picnicking with friends rescues child after child from the nearby river until he finally decides to walk upstream along the river to find out who is pushing the children into the river. (Harris, 1998).

Attachment disorder is an alarming problem. Those who work in the mental health or child welfare fields are well aware that today's children have more serious problems than they did 10 to 20 years ago. This problem has serious repercussions. It is not enough to treat the individuals with this problem, the causes must be explored and prevented. Society cannot afford to ignore the repercussions. Nothing short of a coordinated community approach will suffice. The future of children and society depend upon the community's dedication to this task.

REFERENCES

Bowlby, J. (1982). *Attachment and loss: Attachment* (2nd ed., Vol. 1). New York: Basic Books.

Carnegie Corporation of New York (1994). *Starting points: Meeting the needs of our youngest children.* Waldorf, MD: Carnegie Corporation of New York.

Children's Defense Fund (1998) *The state of America's children, yearbook 1998.* Washington, D.C.: Children's Defense Fund.

Cline, F. W. (1995). *Conscienceless acts societal mayhem.* Golden, CO: The Love and Logic Press.

Harris, I. B. (1998). Pushing kids into the river. *Zero to Three, 18*(5), 1–4.

Karr-Morse, R. & Wiley, M. S. (1997). *Ghosts from the nursery—Tracing the roots of violence.* New York: The Atlantic Monthly Press.

Keck, G. C., & Kupecky, R. M. (1995). *Adopting the hurt child.* Colorado Springs, CO: Pinion Press.

McKelvey, C. A. (Ed.) (1995). *Give them roots, then let them fly: Understanding attachment therapy.* Evergreen, CO: The Attachment Center at Evergreen.

Petit, M. R., & Curtis, P. A. (1997). *Child abuse and neglect: A look at the states: 1997 CWLA stat book.* Washington, D.C.: CWLA Press.

Randolph, L. (1997) *Attachments.* Evergreen, Co: The Attachment Center Press.

Verny, T. & Kelly, J. (1981) *The secret life of the unborn child.* New York: Dell.

Verrier, N. N. (1994). *The primal wound: Understanding the adopted child.* Baltimore, MD: Gateway Press.

Zuckerman, B., & Beardslee, W. (1987). Maternal depression: An issue for pediatricians. *Pediatrics, 79,* 110–117.

INDEX